Carbohydrates:

The Sweet Molecules of Life

I do not derive much pleasure from theoretical things. The occupation with the Walden inversion was rather a digression and recuperation from the extensive work on proteins. Moreover, so many limited heads have now jumped on this question, that the delight thereon is spoiled completely.

Emil Fischer, 1912

Carbohydrates:
The Sweet Molecules of Life

Robert V. Stick

Department of Chemistry
The University of Western Australia
35 Stirling Hwy
Crawley
Western Australia 6009
Australia

ACADEMIC PRESS

A Harcourt Science and
Technology Company

San Diego San Francisco New York
Boston London Sydney Tokyo

Copyright © 2001 by ACADEMIC PRESS

Academic Press
A Harcourt Science and Technology Company
Harcourt Place, 32 Jamestown Road, London NW1 7BY, UK
http://www.academicpress.com

Academic Press
A Harcourt Science and Technology Company
525 B Street, Suite 1900, San Diego, California 92101-4495, USA
http://www.academicpress.com

ISBN 0-12-670960-2

A catalogue record for this book is available from the British Library

Typeset by Mathematical Composition Setters Ltd, Salisbury, UK
Printed and bound in Great Britain by MPG Books Ltd, Bodmin

01 02 03 04 05 06 MP 9 8 7 6 5 4 3 2 1

Contents

Preface and Acknowledgements

To me, there seem to be only two reasons for writing a book. The first is to disseminate new knowledge and, for many branches of science, this is better done through the plethora of scientific journals that exist today. The second reason is to deliver a new treatment of a subject and that, precisely, is what this book sets out to do.

Carbohydrates are mentioned or implied in a household context every day — "pass the sugar, please", "you won't have any energy if you don't eat properly", "you need to have more fibre in your diet", "I hear he's suffering from the sugar" (diabetes) and so on it goes. In fact, for decades, carbohydrates were simply viewed as the powerhouse that provided the energy to drive the many biochemical processes that keep us going. Carbohydrates lived in the shadow of two other great biomolecules, proteins and nucleic acids, until scientists realized the connection between the structural diversity of carbohydrates and their role in a whole range of biochemical processes.

Today, carbohydrates are implicated in intercellular recognition, bacterial and viral infection processes, the fine tuning of protein structure, the inflammation event and some aspects of cancer, to name but a few. This broadening of carbohydrate activity has caused a renaissance in structure determination and synthetic activity, so much so that some of the top chemists and biochemists in the world have been attracted to this area of intractable "gums and syrups", previously the domain of those strange, misguided people called "sugar chemists".

This book, then, will tell you all about carbohydrates. It will give the basic knowledge about the subject, bound together with some of the history and feeling of the times. What was it really like in Emil Fischer's laboratory in the late 1800s? Who followed in the great man's footsteps, who are the emerging giants of carbohydrate chemistry? When a subject is too large or demanding to be treated in the depth that this book allows, pertinent references will be given to aid the reader. A general comment on the selection of references: when deemed appropriate, the reference to an original piece of work will be given; otherwise, use will be made of a modern review article or a recent paper which nicely summarizes the area.

All in all, this is a modern book about an old subject, but one which continues to show more of its true self as the years pass by – I enjoyed writing it, I hope that you will enjoy reading it!

The book presumes that the reader will have a knowledge of general organic chemistry, probably to the second year level, but requires no background in carbohydrates. The strength of the book is synthesis, ultimately that of the bond which holds two sugar residues together. Towards the end, when the demands of size and subject matter authority were coming into play, an effort was made to introduce pertinent aspects of "glycobiology", the role of carbohydrates in the world of biology. However, the author stresses the need to consult other works to gain any real knowledge about glycobiology and related subjects – indeed, the text by Lehmann (see the Appendix) would be an excellent adjunct to the book here.

Sheri Harbour typed the entire manuscript and it was read, with many suggestions for improvement, by the "Elm Street Boys", David Vocadlo and Spencer Williams, and Steve Withers, Bruce Stone, John Stevens and Matthew Tilbrook. Steve Withers and the Department of Chemistry at the University of British Columbia were my patrons during the writing and Frieder Lichtenthaler kindly helped with the photographs of Fischer. To all of these people, my sincere thanks.

Robert V. Stick

Abbreviations

Ac	acetyl
AIBN	2,2′-azobisisobutyronitrile
All	allyl (prop-2-enyl)
Ar	aryl
BMS	*tert*-butyldimethylsilyl
Bn	benzyl (phenylmethyl)
BPS	*tert*-butyldiphenylsilyl
Bz	benzoyl
CAN	cerium(IV) ammonium nitrate
ClAc	chloroacetyl
CMP	cytidine 5′-monophosphate
CSA	camphor-10-sulfonic acid
cy	cyclohexyl
DABCO	1,4-diazabicyclo[2.2.2]octane
DAST	diethylaminosulfur trifluoride
DBU	1,5-diazabicyclo[5.4.0]undec-5-ene
DCC	dicyclohexylcarbodiimide
DCE	1,2-dichloroethane
DDQ	2,3-dichloro-5,6-dicyanobenzoquinone
DEAD	diethyl azodicarboxylate
DIAD	diisopropyl azodicarboxylate
DMAP	4-(dimethylamino)pyridine
DMDO	dimethyldioxirane
DME	1,2-dimethoxyethane
DMF	dimethylformamide
DMSO	dimethyl sulfoxide
DMTST	dimethyl(methylthio)sulfonium triflate
DNP	2,4-dinitrophenyl
DTBMP	2,6-di-*tert*-butyl-4-methylpyridine
DTBP	2,6-di-*tert*-butylpyridine
Fmoc	fluorenylmethylenoxycarbonyl

GDP	guanosine 5′-diphosphate
HMPA	hexamethylphosphoramide
IDC	iodonium dicollidine
Im	1-imidazyl
LDA	lithium diisopropylamide
MCPBA	3-chloroperbenzoic acid
MNO	4-methylmorpholine *N*-oxide
ms	molecular sieves
Ms	mesyl (methanesulfonyl)
NBS	*N*-bromosuccinimide
NIS	*N*-iodosuccinimide
PCC	pyridinium chlorochromate
PDC	pyridinium dichromate
PEG	poly(ethylene glycol)
Ph	phenyl
Phth	phthaloyl
Piv	pivaloyl (2,2-dimethylpropanoyl)
PMB	4-methoxybenzyl
PNP	4-nitrophenyl
PTSA	4-toluenesulfonic acid
py	pyridine
rt	room temperature
SF	Selectfluor {1-chloromethyl-4-fluoro-1,4-diazoniabicyclo[2.2.2]octane bis(tetrafluoroborate)}
TCP	tetrachlorophthaloyl
TEMPO	2,2,6,6-tetramethylpiperidine-1-oxyl
Tf	triflyl (trifluoromethanesulfonyl)
THF	tetrahydrofuran
THP	tetrahydropyranyl
TIPS	triisopropylsilyl
TMP	2,2,6,6-tetramethylpiperidide
Tol	tolyl (4-methylphenyl)
TPAP	tetrapropylammonium perruthenate
Tr	trityl (triphenylmethyl)
Ts	tosyl (4-toluenesulfonyl)
UDP	uridine 5′-diphosphate
UTP	uridine 5′-triphosphate

Chapter 1

The Meaning of Life

Do you ever contemplate your existence? Is it not a marvel to think that you started out as a few cells which, so far, have culminated in where you are today? How did you survive those years of infancy, virtually dependent on the skills and protective nature of your parents, followed by those teenage and young adult years when so many activities were, in retrospect, life threatening? While all of this was going on at the macroscopic level, similar wonderments were occurring inside you. Molecules were being broken down, other molecules were being assembled. Your DNA was being faithfully copied, with little error – even then, some helper molecule would come along and repair the damage. The proteins of your body were being assembled and some of these, the enzymes, carried out miraculous chemical transformations rapidly and specifically. Infection was recognized, a defence mounted and the harmful organism finally conquered and ousted. Broken and damaged parts were, occasionally with outside help, made to mend beautifully. To top this all off, finely tuned biochemical pathways provided the energy to drive all of these events. At the hub, carbohydrates!

Well, what exactly is a carbohydrate? As the name implies, an empirical formula of $C.H_2O$ (or CH_2O) was often encountered, with molecular formulae of $C_5H_{10}O_5$ and $C_6H_{12}O_6$ being most common. The water solubility of these molecules was commensurate with the presence of hydroxyl groups and there was always evidence for the carbonyl group of an aldehyde or ketone. These polyhydroxylated aldehydes and ketones were termed *aldoses* and *ketoses*, respectively – for the more common members, actually, *aldopentoses/aldohexoses* and *ketopentoses/ketohexoses*. Very early on, it became apparent that larger molecules existed that could be converted, by hydrolysis, into the smaller and more common units – *monosaccharides* from *polysaccharides*. Nowadays, the definition of what is a carbohydrate has been much expanded to include oxidized or reduced molecules and those that contain other types of atoms (often nitrogen). The term "sugar" is used to describe the monosaccharides and the somewhat higher molecular weight disaccharides, trisaccharides and so on.

Chapter 2

The Early Years

The "man amongst men" in the late 1800s was, undoubtedly, Emil Hermann Fischer.[a] To try to appreciate the genius and elegance of Fischer's work with sugars, let us consider the conditions and resources available in a typical laboratory in Germany in those days. The photograph (Figure 1) of von Baeyer's research group in Munich in 1878 speaks volumes.

Fischer, appropriately seated next to von Baeyer,[b] is surrounded by formally attired, austere men, some wearing hats (for warmth?) and many sporting a beard or moustache. The large hood in the background carries an assortment of apparatus, presumably for the purpose of microanalysis.

Microanalysis, performed meticulously by hand, was the cornerstone of Fischer's work on sugars. Melting point and optical rotation were essential adjuncts in the determination of chemical structure and equivalence. All of this required *pure* chemical compounds, necessitating crystallinity at every possible opportunity — sugar "syrups" decomposed on distillation and the concept of chromatography was barely embryonic in the brains of Day[c] and Tswett.[d] Fortunately, many of the naturally occurring sugars were found to be crystalline; however, upon chemical modification, their products often were not. This, coupled together with the need to investigate the chemical structure of sugars, encouraged Fischer and others to invoke some of the simple reactions of organic chemistry, and to invent new ones.

---

[a] Emil Hermann Fischer (1852–1919), PhD (1874) under von Baeyer at the University of Strassburg, professorships at Munich, Erlangen (1882), Würzburg (1885) and Berlin (1892). Nobel Prize (1902).

[b] Johann Friedrich Wilhelm Adolf von Baeyer (1835–1917), PhD under Kekulé and Hofmann at the Universities of Heidelberg and Berlin, respectively, professorships at Strassburg and Munich. Nobel Prize (1905).

[c] David Talbot Day (1859–1925), PhD at the Johns Hopkins University, Baltimore (1884), chemist, geologist and mining engineer.

[d] Mikhail Semenovich Tswett (1872–1919), DSc at the University of Geneva, Switzerland (1896), chemist and botanist.

Figure 1 Photograph of the Baeyer group in 1878 at the laboratory of the University of Munich (room for combustion analysis), with inscriptions from Fischer's hand. This, and the photograph on page 17, of which the originals are in the "Collection of Emil Fischer Papers" (Bancroft Library, University of California, Berkeley), were obtained from Professor Frieder W. Lichtenthaler (Darmstadt, Germany), who used them in his article (*Angew. Chem. Int. Ed. Engl.* 1992, **31**, 1541).

Oxidation was an operationally simple task for the early German chemists. The aldoses, apart from showing the normal attributes of a *reducing sugar* (forming a beautiful silver mirror when treated with Tollens'[e] reagent or causing the precipitation of brick-red cuprous oxide when subjected to Fehling's[f]

---

[e] Bernhard C. G. Tollens (1841–1918), professor at the University of Göttingen.

[f] Hermann von Fehling (1812–1885), professor at the University of Stuttgart.

solution), were easily oxidized by bromine water to carboxylic acids, termed *aldonic acids*.

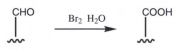

Moreover, heating of the newly formed aldonic acid often formed cyclic esters, lactones.

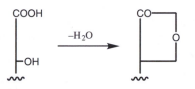

Ketoses, not surprisingly, were not oxidized by bromine water and could thus be distinguished simply from aldoses.

Dilute nitric acid was also used for the oxidation of aldoses, this time to dicarboxylic acids, termed *aldaric acids*.

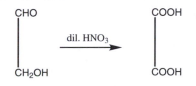

Lactone formation from these diacids was still observed, with the formation of more than one lactone not being uncommon.

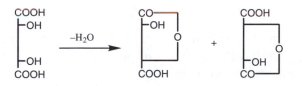

Reduction of sugars was most conveniently performed with sodium amalgam (NaHg) in ethanol. Aldoses yielded one unique *alditol* whereas ketoses, for reasons that may already be apparent, gave a mixture of two alditols:

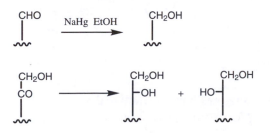

Fischer, with interests in chemicals other than carbohydrates, treated a solution of the benzenediazonium ion (the cornerstone of the German dye-stuffs industry) with sulfur dioxide and, in so doing, discovered phenylhydrazine (1875):

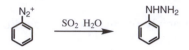

Fischer soon found that phenylhydrazine was useful for the characterization of the somewhat unreliable sugar acids by converting them into their very crystalline phenylhydrazides:

Phenylhydrazine also transformed aldehydes and ketones into phenylhydrazones and, not remarkably, similar transformations were possible with aldoses and ketoses:

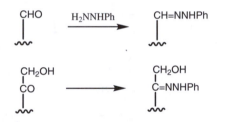

The remarkable aspect of this work was that both aldoses and ketoses, when treated more vigorously with an *excess* of phenylhydrazine, were converted into unique derivatives, **phenylosazones**:

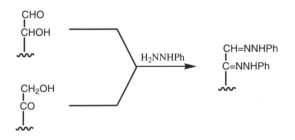

The different phenylosazones had distinctive crystalline forms and, as well, were formed at different rates from the various parent sugars.

Another carbohydrate chemist of the times, Kiliani,[g] amply acknowledged by Fischer but generally underrated by his peers, had applied some well-known chemistry to aldoses and ketoses, namely the addition of hydrogen cyanide. The products, after acid hydrolysis, were aldonic acids. Fischer took the lactones derived from these acids and showed that they could be reduced to aldoses, containing an extra carbon atom:

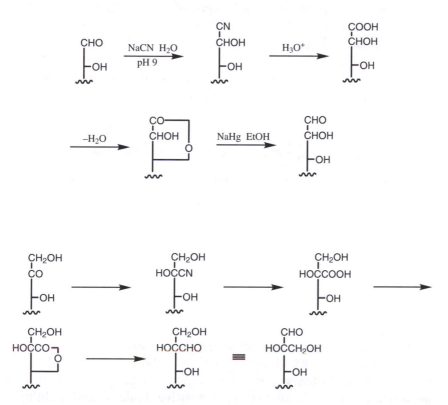

and

Not so obviously, this synthesis converts an aldose or ketose into *two* new aldoses. Fischer used and developed this **ascent** (adding one carbon) of the homologous aldose series so well that it is known as the Kiliani–Fischer synthesis.

It was logical that, if "man" could ascend the aldose series, "he" should also be able to descend the series — so were developed various methods for this **descent**. Perhaps the most well known is that devised by Ruff[h] — the aldose is

---

[g] Heinrich Kiliani (1855–1945), PhD under Erlenmeyer and von Baeyer, professor at the University of Freiburg.

[h] Otto Ruff (1871–1939), professorships at Danzig and Breslau.

first oxidized to the aldonic acid and subsequent treatment of the calcium salt of the acid with hydrogen peroxide gives the aldose:

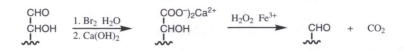

It is an interesting complement to the ascent of a series that the (Ruff) descent converts two aldoses into just one new aldose.

The final transformation that was available to Fischer, albeit somewhat late in the piece, was of an informational, rather than a preparative, nature. Lobry de Bruyn and Alberda van Ekenstein[1,2] announced the **rearrangement** of aldoses and ketoses upon treatment with dilute alkali:

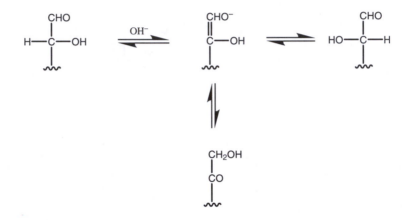

This simple, enolate-driven sequence allowed the isomerization of one aldose into its C2 epimer, together with the formation of the structurally related ketose. It also explained the observation that ketoses, although not oxidizable by bromine water (at a pH below 7), gave positive Tollens' and Fehling's tests (conducted with each reagent under alkaline conditions).

Fischer now had the necessary chemical tools (and intellect!) to launch an assault on the structure determination of the carbohydrates.

References

1.   Lobry de Bruyn, C. A. and Alberda van Ekenstein, W. (1895). *Recl. Trav. Chim. Pays-Bas*, **14**, 203.
2.   Speck, J. C. Jr (1958). *Adv. Carbohydr. Chem.*, **13**, 63.

Chapter 3

"Grandfather Glucose"[a]

(+)-Glucose from a variety of sources (fruits and honey), (+)-galactose from the hydrolysis of "milk sugar" (lactose), (−)-fructose from honey, (+)-mannitol from various plants and algae, (+)-xylose and (+)-arabinose from the acid treatment of wood and beet pulp, respectively — these were the sugars available to Fischer when he started his seminal studies on the structures of the carbohydrates in 1884 in Munich.

What were the established facts about (+)-glucose at that time? It was a reducing sugar that could be oxidized to gluconic acid with bromine water and to glucaric acid with dilute nitric acid. That the six carbon atoms were in a contiguous chain had been shown by Kiliani; the conversion of (+)-glucose into a mixture of heptonic acids (by conventional Kiliani extension), followed by treatment of this mixture with red phosphorus and hydrogen iodide (strongly reducing conditions), gave heptanoic acid:

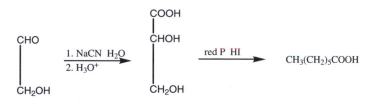

Thus, the structure of (+)-glucose was established as a straight-chain, polyhydroxylated aldehyde:[b]

---

[a] This was part of a title used by Professor Bert Fraser-Reid in an article (*Acc. Chem. Res.*, 1996, **29**, 57) describing some of his work.

[b] A similar sequence on (−)-fructose produced 2-methylhexanoic acid, establishing the fact that

The theories of Le Bel and van't Hoff, around 1874, decreed that a carbon atom substituted by four different groups (as we have with the sugars) should be tetrahedral in shape and able to exist as two separate forms, non-super-imposable mirror images and, thus, isomers. These revolutionary ideas were seized upon and endorsed by Fischer and formed the cornerstone for his arguments on the structure of (+)-glucose.

In order to simplify the ensuing arguments, let us digress to the simplest aldose, the aldotriose, glyceraldehyde (formaldehyde and glycolaldehyde, although formally sugars, are not regarded as such):

The two isomers, in fact enantiomers, may be represented using Fischer projection formulae:[c]

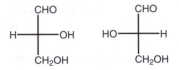

Rosanoff, an American chemist of the times, decreed, quite arbitrarily, that (+)-glyceraldehyde would be represented by the first of the two enantiomers and its unique *absolute configuration* was described a little later by the use of the small capital letter, D:[1,2]

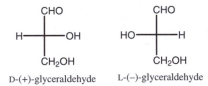

D-(+)-glyceraldehyde   L-(−)-glyceraldehyde

(−)-fructose was a 2-keto sugar:

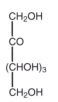

[c] Such formulae were first announced by Fischer in 1891 and, besides simplifying the depiction of the sugars, were universally accepted. Being planar projections, the actual stereochemical information is available only if you know the "rules" — horizontal lines actually represent bonds above the plane, vertical lines those bonds below the plane. Only one "operation" is hence allowed with Fischer projection formulae — a rotation of 180° in the plane.

Fischer, in an effort to thread together the jumble of experimental results on the sugars, had earlier decided that (+)-glucose would be drawn with the hydroxyl group to the right at its bottom-most (highest numbered) "substituted" carbon {the same absolute configuration as (+)-glyceraldehyde}:

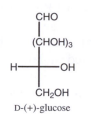

D-(+)-glucose

The challenge remaining was to elucidate the *relative configuration* of the other three centres (eight possibilities)!

What follows is an account of Fischer's elucidation of the structure of D-(+)-glucose, interspersed with anecdotal information gleaned from a wonderful article by Professor Frieder Lichtenthaler (Darmstadt, Germany),[3] to celebrate the announcement of the structure of (+)-glucose in 1891.[4,5] To begin, a passage from a letter by Fischer to von Baeyer:

> *The investigations on sugars are proceeding very gradually. It will perhaps interest you that mannose is the geometrical isomer of grape sugar. Unfortunately, the experimental difficulties in this group are so great, that a single experiment takes more time in weeks than other classes of compounds take in hours, so only very rarely a student is found who can be used for this work. Thus, nowadays, I often face difficulties in trying to find themes for the doctoral theses.*

On top of this "soul searching" by Fischer, consider the following experimental results:

$$\text{D-xylose} \xrightarrow{\text{NaHg EtOH}} \text{xylitol} \qquad [\alpha]_D 0°$$

$$\text{L-arabinose} \xrightarrow{\text{NaHg EtOH}} \text{arabinitol} \quad [\alpha]_D 0°$$

Both would appear to be achiral (meso) compounds, but what of:

$$\text{D-xylose} \xrightarrow{\text{HNO}_3} \text{xylaric acid} \qquad [\alpha]_D 0°$$

$$\text{L-arabinose} \xrightarrow{\text{HNO}_3} \text{arabinaric acid} \quad [\alpha]_D -22.7°$$

In the two sets of experiments, the "ends" of the sugar chains were identical (both "CH$_2$OH" or both "COOH"). Clearly, the xylitol and xylaric acid were meso compounds, but the arabinaric acid was not! This meant that the arabinitol *had* to be chiral; only in the presence of borax (which forms complexes with polyols) was Fischer able to obtain a very small, negative rotation for arabinitol. Would we, in this day and age, be so careful and observant?

To the proof of the structure of (+)-glucose:

1. Because Fischer had arbitrarily placed the hydroxyl group at C5 on the right for (+)-glucose, all interrelated sugars must have the same (D-) absolute configuration.
2. Arabinose, on Kiliani–Fischer ascent, gave a mixture of glucose and mannose.[d]

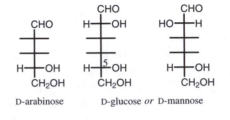

D-arabinose   D-glucose *or* D-mannose

---

[d] Mannose was first prepared (1887) in very low yield by the *careful* (HNO_3) oxidation of mannitol and later obtained from the acid hydrolysis of "mannan" (a polysaccharide) present in tagua palm seeds (ivory nut). That glucose and mannose were epimers at C2 was shown by the following transformations:

glucose ⟶ glucose phenylosazone ⟵ mannose

        ⟶ phenylhydrazone   phenylhydrazone ⟵
           mp 144–145°C   ——   mp 188°C

L-Glucose, together with L-mannose, had been prepared earlier by Kiliani–Fischer extension of (+)-arabinose (actually L-arabinose) from sugar beet:

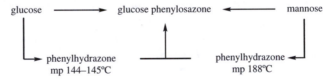

D-Gulose, together with D-idose, arose when (−)-xylose (actually D-xylose) from cherry gum was subjected to a Kiliani–Fischer synthesis:

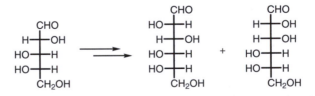

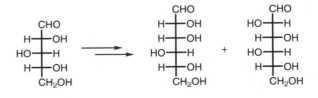

3. Arabinaric acid was *not* a meso compound and, therefore, the hydroxyl group at C2 of D-arabinose *must* be to the left.

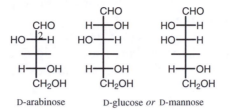

D-arabinose   D-glucose *or* D-mannose

4. *Both* glucaric and mannaric acids are optically active — this places the hydroxyl group at C4 of the hexoses on the right.[e]

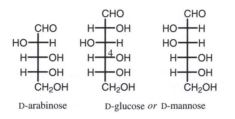

D-arabinose   D-glucose *or* D-mannose

5. D-Glucaric acid comes from the oxidation of D-glucose but L-glucaric acid can be obtained from L-glucose *or* D-gulose.[d] This is only possible if D-gulose is related to L-glucose by a "head to tail" swap:

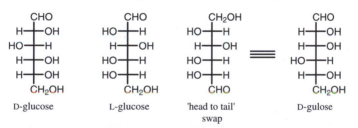

D-glucose   L-glucose   'head to tail'   D-gulose
                                        swap

This wonderful piece of analysis thus provided unequivocal structures for three (of the possible eight) D-aldohexoses and one (of the possible four) 2-keto-D-hexose:[f]

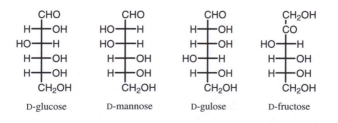

D-glucose   D-mannose   D-gulose   D-fructose

---

[e] The relative configuration of D-arabinose is now established.

[f] Glucose and fructose (and for that matter, mannose) gave the same phenylosazone and were interrelated products of the Lobry de Bruyn–Alberda van Ekenstein rearrangement.

After the elucidation of the structure of D-arabinose and the four D-hexoses above, it was but a "simple" matter, employing similar chemical transformations and logic, to unravel the structure of D-galactose; Kiliani, in 1888, had secured the structure of D-xylose.

These six aldoses and one ketose are but members of the sugar "family trees", with glyceraldehyde at the base for the aldoses and dihydroxyacetone for the 2-ketoses (Figures 2 and 3). There are various interesting aspects of these family trees:

- The trees are constructed systematically, *viz*. hydroxyl groups are placed to the "right" (R) or "left" (L) according to the designation in the left-side margin.
- When applied to this system, the various mnemonics[g] enable one to write the structure of any named sugar or, in the reverse, to name any sugar structure.
- As Fischer encountered unnatural sugars through synthesis, additional names had to be found; thus, "lyxose" is an anagram of "xylose", "gulose" is an abbreviation/rearrangement of "glucose".
- It is well worthwhile to consider the simple name, D-glucose; it describes a unique molecule with four stereogenic centres and *must* be superior to the more modern (2R, 3S, 4R, 5R)-2,3,4,5,6-pentahydroxyhexanal.[h]

It was not until 1951 that the D-absolute configuration for (+)-glucose, arbitrarily chosen by Fischer some 75 years earlier, was proven to be correct. By a series of chain degradations, (+)-glucose was converted into (−)-arabinose and then (−)-erythrose. Chain extension of (+)-glyceraldehyde also gave (−)-erythrose, together with (−)-threose. Oxidation of (−)-threose gave (−)-tartaric acid, the enantiomer of (+)-tartaric acid.

(+)-Tartaric acid had been converted independently into a beautifully crystalline sodium/rubidium salt; an X-ray structure determination of this salt

---

[g] Figure 2:
the tetroses — "ET" (the film)
the pentoses — "raxl" is perhaps less flowery
the hexoses — designed by Louis and Mary Fieser (Harvard University)

Figure 3:
dihydroxyacetone — an achiral molecule
the term "ulose" is formal nomenclature for a ketose.

[h] The only other bastion of the D/L system is that of amino acids; however, with the difficulty in defining what an L-amino acid (with generally just one stereogenic centre) actually is, any self-respecting scientist should revert to the more modern, and unambiguous, *R/S* system.

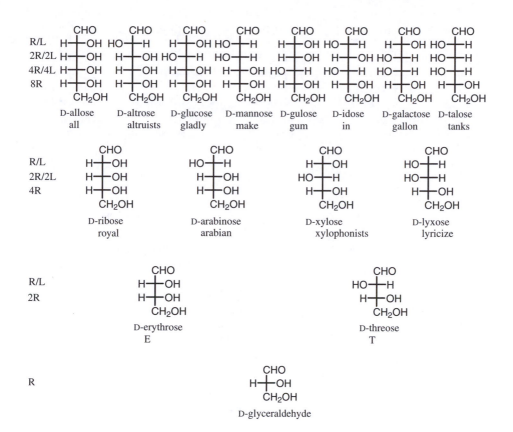

Figure 2 The D- family tree of the aldoses.

showed it to have the following absolute configuration:[6]

This defined the structures of (+)-tartaric acid, (−)-tartaric acid and (−)-threose as

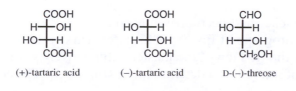

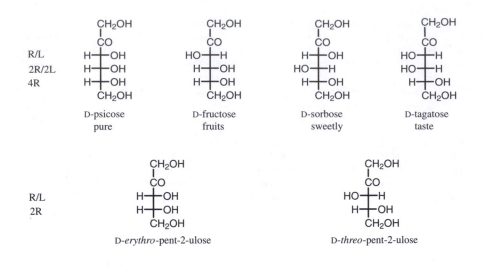

Figure 3 The D- family tree of the ketoses.

This allowed the assignment of absolute configuration to (−)-erythrose and (+)-glyceraldehyde:

Rosanoff and Fischer had been proven correct.

Finally, a photograph of Fischer in his later years at the University of Berlin is exceptional in that it shows "the master" still actively working at the bench, with a face full of interest and determination (Figure 4). The constant exposure

Figure 4 Emil Fischer around the turn of the century in his laboratory at the University of Berlin. This, and the photograph on page 4, of which the originals are in the "Collection of Emil Fischer Papers" (Bancroft Library, University of California, Berkeley), were obtained from Professor Frieder W. Lichtenthaler (Darmstadt, Germany), who used them in his article (*Angew. Chem. Int. Ed. Engl.* 1992, **31**, 1541).

to chemicals, particularly phenylhydrazine (osazone formation) and mercury (NaHg reductions), caused chronic poisoning and eczema and, coupled with the loss of two of his three sons in World War I, Fischer took his own life in 1919.

References

1. Rosanoff, M. A. (1906). *J. Am. Chem. Soc.*, **28**, 114.
2. Hudson, C. S. (1948). *Adv. Carbohydr. Chem.*, **3**, 1.
3. Lichtenthaler, F. W. (1992). *Angew. Chem. Int. Ed. Engl.*, **31**, 1541.
4. Fischer, E. (1891). *Ber. Dtsch. Chem. Ges.*, **24**, 1836.
5. Fischer, E. (1891). *Ber. Dtsch. Chem. Ges.*, **24**, 2683.
6. Bijvoet, J. M., Peerdeman, A. F. and van Bommel, A. J. (1951). *Nature*, **168**, 271.

Chapter 4

Heidi und Heinz

Let me now set the scene to 1891 where Fischer, after his announcement of the structure of (+)-glucose, is resting, for once, rather contentedly in his office. Suddenly, in through the door, there burst two of his prize pupils, Heidi and Heinz. Heidi declares:

> *Sehr geehrter Herr Professor Doktor Fischer! Ich habe den Trauben-zucker völlig purifizieren können. Er hat einen Schmelzpunkt von hundert sechs und vierzig Grad und eine Rotation von plus hundert zwölf Grad (im Wasser).*

Heinz follows:

> *Sehr geehrter Herr Professor Doktor Fischer! Das ist nicht wahr, was Heidi sagt! Ich habe den Traubenzucker völlig purifizieren können. Er hat einen Schemlzpunkt von hundert fünfzig Grad und eine Rotation von plus neunzehn Grad (im Wasser).*

What is this — *two* forms of pure (+)-glucose, with *almost* the same melting point, but having vastly different values for the specific optical rotation (+112° and +19°)? How could this be so?

So far, we have represented the structure of (+)-glucose as a Fischer projection, a useful convention. However, in real life, as either a solid or in solution, (+)-glucose has a molecular structure which may take up an infinite number of shapes, or *conformations*. If one makes a molecular model of (+)-glucose, a linear, zig-zag conformation seems attractive:

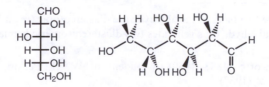

Playing around with this linear conformation, by rotation around the various carbon–carbon bonds present, does *nothing* to the configuration of the molecule but leads to an infinite number of other conformations. One of these conformations, on close scrutiny, has the hydroxyl group on carbon-5 adjacent to the aldehyde group (carbon-1). What follows is a chemical reaction, the nucleophilic addition of the carbon-5 hydroxyl group to the aldehyde group to generate a *hemiacetal*:

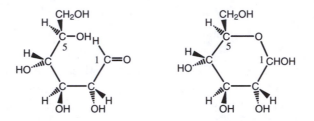

This new chemical structure possesses an extra stereogenic centre (carbon-1) and so the product of the cyclization may exist in two, discrete, isomeric forms:[a]

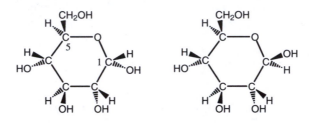

These new cyclic forms of (+)-glucose explained the claims by Heidi and Heinz — indeed, such cyclic forms had been suggested by von Baeyer in 1870 and again by Tollens in 1883.[1] Fischer, somewhat surprisingly, was never completely accepting of these structures. Again, it must be emphasized that the above depictions are each a form of (+)-glucose; carbon-5 still is of the D-absolute configuration and carbons two to four complete the description of D-(+)-glucose.

It will be obvious even at this stage that, to the sugar chemists of the 1900s, communicating with structures anything like the above was a tiresome process — a shorthand had to be developed. In 1926, an eminent chemist of the day, W. N. Haworth,[b] made suggestions towards the 6-membered ring being represented

[a] Put more formally, the two faces of the aldehyde are diastereotopic (*re* and *si*) — addition of the hydroxyl group to the aldehyde thus generates two diastereoisomeric hemiacetals, not necessarily in equal amounts.

[b] Walter Norman Haworth (1883–1950), a student of W. H. Perkin, PhD under Wallach (Göttingen). Nobel Prize (1937).

as a hexagon with the front edges emboldened, causing the hexagon to be viewed front-edge-on to the paper:[2]

The two remaining bonds to each carbon are depicted, one above and one below the plane of the hexagon. Now, the two cyclic forms of (+)-glucose can be drawn swiftly and accurately:[c,d]

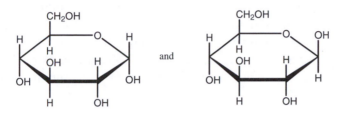

Some years earlier (1913),[4] complexation studies with boric acid had shown that the more highly rotating isomer of (+)-glucose ($[\alpha]_D$ +112°) possessed a *cis*-relationship between the hydroxyl groups at carbons one and two. Full structural assignments were now possible and, to simplify the matter of

---

[c] For an excellent discussion on the "rotational operations" allowed with Haworth formulae, see "Advanced Sugar Chemistry: Principles of sugar stereochemistry" by R. S. Shallenberger (AVI Publishing Company Inc., Westport, Connecticut; 1982, p. 110) — "Haworth structures can be rotated *on* the plane of the paper on which they are drawn if, and only if, the identity of the leading edge of the structure is not lost".

[d] Another convention, suggested by John A. Mills in 1955 and still in some use,[3] again uses a hexagon, but to be viewed in the plane of the paper. Nowadays, hydrogen substituents are not shown, but others are, using "wedge" (above) or "dash" (below) notation. An extension of the Mills' convention can be used to indicate a racemic structure, with the relative configuration still obvious:

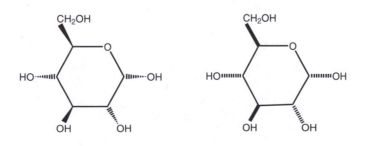

communication even further, formal names were given to the two isomers:

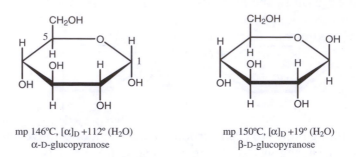

mp 146°C, [α]$_D$ +112° (H$_2$O)
α-D-glucopyranose

mp 150°C, [α]$_D$ +19° (H$_2$O)
β-D-glucopyranose

The term "ose" still indicates a sugar, and "pyranose" a sugar having a six-membered cyclic structure.[e] The terms "α" and "β" refer to the particular *anomer* (diastereoisomer, epimer) and carbon-1, for aldoses, is the *anomeric carbon*.

Let us return to this question of the shape of the (+)-glucose chain and consider what happens when we encounter a different conformation, one where the hydroxyl group on carbon-*4* finds itself adjacent to the aldehyde group:

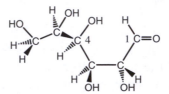

Again, hemiacetal formation is possible, resulting in the formation of two new anomers (logically drawn according to the Haworth convention):

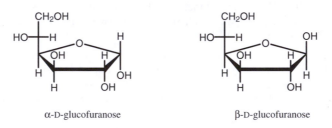

α-D-glucofuranose

β-D-glucofuranose

---

[e] By analogy with the molecule, pyran:

There are few data available for these "furanose"[f] forms of D-glucose, simply because they have never been isolated; normal, crystalline (+)-glucose is either one of the pure pyranose anomers, or a mixture thereof.

Before proceeding any further, it is worth looking at a few sugars other than D-glucose and considering their cyclic structures:

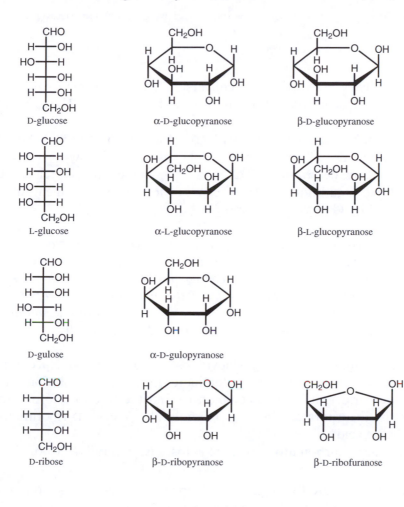

<hr>

[f] From the molecule, furan:

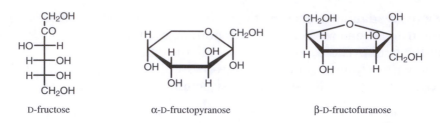

D-fructose   α-D-fructopyranose   β-D-fructofuranose

Several pertinent points emerge:

- In the Fischer/Haworth "interconversion", all hydroxyl groups on the "right" in a Fischer projection are placed "below" in the Haworth, and all those on the "left" are "above".
- In a D-aldohexose, the "CH_2OH" group at carbon-5 is "above" in the Haworth pyranose form; in the L-aldohexose, it is "below".
- The anomeric descriptions, "α" and "β", are obviously related to the absolute configuration — there can be no clearer statement than that enunciated by Collins and Ferrier:[5]

> *For D-glucose and all compounds of the D-series, α-anomers have the hydroxyl group at the anomeric centre projecting downwards in Haworth formulae; α-L-compounds have this group projecting upwards. The β-anomers have the opposite configurations at the anomeric centre, i.e. the hydroxyl group projects upwards and downwards for β-D- and β-L-compounds, respectively.*

Thus, the enantiomer of α-D-glucopyranose, as shown, is α-L-glucopyranose.[g]
- Whereas the pyranose forms dominate in aqueous solution of most monosaccharides, it is quite common to find the furanose form when the sugar is incorporated into a biomolecule, *e.g.* β-D-ribofuranose in ribonucleic acid.
- The anomeric carbon atom of the 2-ketoses is, naturally, carbon-2.

The cyclic structure for sugars now helped to explain several observations that had been made by the German pioneers in the nineteenth century:

- Aldoses do not form addition compounds with sodium bisulfite and fail to give some of the very sensitive and characteristic colour tests for aldehydes.

---

[g] Originally, another famous carbohydrate chemist, Claude S. Hudson (1881–1952, a student of van't Hoff, PhD at Princeton University) proposed a definition for "α" and "β" based on the relative magnitude of the specific optical rotation.[6]

- Aldoses, generally, tended to react with hydrogen cyanide and with phenylhydrazine more slowly than do normal aldehydes.
- With careful choice of reagents and conditions, (+)-glucose could be converted into two different pentaacetates:

penta-*O*-acetyl-α-D-glucopyranose   penta-*O*-acetyl-β-D-glucopyranose

Still, aldoses did *eventually* show most of the reactions characteristic of an aldehyde – how then to explain this apparent dichotomy? Let us return for one last look at our "Heidi und Heinz" pantomime. Both of our young researchers, somewhat crestfallen, have returned to the Professor's office to announce (Heidi) that the specific optical rotation of one form of (+)-glucose has fallen from +112° to +52° and (Heinz, almost in tears) that of the other form has risen from +19° to the same +52°.

So was discovered the phenomenon of **mutarotation** — the change in optical rotation with time. This simple physical observation was easily explained by a consideration of the following equilibrium:

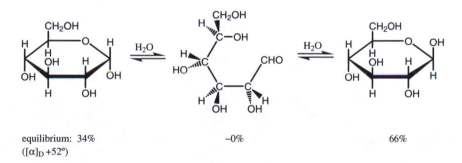

equilibrium: 34%   ~0%   66%
($[\alpha]_D$ +52°)

The value of +52° was obviously the specific optical rotation for the *mixture* at equilibrium. Knowing the values for the two pure anomers allows one to calculate the percentage of each anomer present. In aqueous solutions of (+)-glucose, then, there is virtually none of the acyclic form actually present, but it is always *available* by chemical disruption of the above equilibrium.[h]

---

[h] One final term needs to be introduced here, that of a **"free sugar"** — presumably, a sugar "free" to undergo the process of mutarotation and a term synonymous with "hemiacetal" and "reducing sugar".

Finally, a few words on the actual determination of "ring size" in carbohydrates. Years ago, the classic approaches involved such processes as "methylation analysis"[7-9] and "periodate cleavage".[10-13] Although these methods are still in some use, it is the power of nuclear magnetic resonance spectroscopy, both ^{1}H and ^{13}C, that is brought to bear on such problems nowadays.[14,15]

References

1. Lichtenthaler, F. W. (1992). *Angew. Chem. Int. Ed. Engl.*, **31**, 1541.
2. Drew, H. D. K. and Haworth, W. N. (1926). *J. Chem. Soc.*, 2303.
3. Mills, J. A. (1955). *Adv. Carbohydr. Chem.*, **10**, 1.
4. Böeseken, J. (1949). *Adv. Carbohydr. Chem.*, **4**, 189.
5. Collins, P. M. and Ferrier, R. J. (1995). *Monosaccharides: Their Chemistry and Their Roles in Natural Products*, John Wiley & Sons, Chichester, p. 14.
6. Hudson, C. S. (1909). *J. Am. Chem. Soc.*, **31**, 66.
7. Hirst, E. L. and Purves, C. B. (1923). *J. Chem. Soc. (Trans.)*, **123**, 1352.
8. Hirst, E. L. (1926). *J. Chem. Soc.*, 350.
9. Haworth, W. N., Hirst, E. L. and Learner, A. (1927). *J. Chem. Soc.*, 1040, 2432.
10. Jackson, E. L. and Hudson, C. S. (1937). *J. Am. Chem. Soc.*, **59**, 994.
11. Jackson, E. L. and Hudson, C. S. (1939). *J. Am. Chem. Soc.*, **61**, 959.
12. Maclay, W. D., Hann, R. M. and Hudson, C. S. (1939). *J. Am. Chem. Soc.*, **61**, 1660.
13. Bobbitt, J. M. (1956). *Adv. Carbohydr. Chem.*, **11**, 1.
14. Williams, C. and Allerhand, A. (1977). *Carbohydr. Res.*, **56**, 173.
15. Horton, D. and Walaszek, Z. (1982). *Carbohydr. Res.*, **105**, 145.

Chapter 5

The Shape of Things to Come

A century of investigation had unlocked the stereochemical secrets of D-(+)-glucose, depicted as either a Fischer projection or, more accurately as we have seen, a cyclic molecule in a Haworth formula:[a]

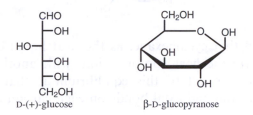

D-(+)-glucose   β-D-glucopyranose

In the early 1900s, most chemists believed that the pyranose ring was non-planar. However, it took the work of Hassel,[b] using electron diffraction studies in the gas phase, to put some substance into these notions; the cyclohexane ring was shown to have a non-planar shape (conformation), actually that of a **chair**:[1]

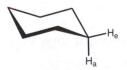

Some years later, Barton[b] recognized the importance of the two different types of bonds present in cyclohexane (**equatorial** and **axial**) and used this information to explain the conformation and reactivity in molecules such as the steroids.[2,3] The beauty of this result was that, in the chair conformation for cyclohexane, each carbon was almost exactly tetrahedral in shape – cyclohexane, as predicted and shown, exhibited no Baeyer "angle strain".

---

[a] From now on, hydrogen atoms bound to carbon will generally not be shown.

[b] Odd Hassel (1897–1981, PhD from the University of Berlin), a Norwegian, shared a Nobel Prize (1969) with Derek Harold Richard Barton (1918–1998), British.

A further advance by Hassel was to predict that the conformation of the pyranose ring would also be non-planar and, probably, again a chair:

For β-D-glucopyranose, the most common monosaccharide in the free form on our planet, all of the hydroxyl substituents on the pyranose ring had to be equatorially disposed (otherwise the molecule is no longer β-D-glucose!), the most stable arrangement possible:

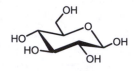

It is well known that cyclohexane, as the neat liquid or in solution, is in rapid equilibrium, *via* the **boat** conformation, with another, but degenerate, chair conformation; a result of this equilibrium is that there is a general interchange of equatorial and axial bonds on each carbon atom:

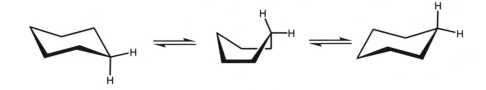

What would be the consequences, if any, of such a process applied to β-D-glucopyranose?

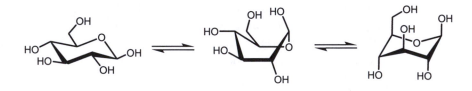

Again an equilibrium is possible, and again *via* a boat conformation, but the new chair conformation is obviously different here from the original — with only axial substituents, the energy of the new conformation is significantly higher (some 25 kJ mol^{-1}).

How, then, do we actually establish the **preferred conformation** for a molecule such as β-D-glucopyranose?

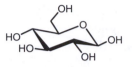

When the molecule in question is crystalline, then a single crystal, X-ray structure determination will yield both the molecular structure and the conformation. When the molecule is a liquid, or in solution, ^{1}H nuclear magnetic resonance spectroscopy will often give the answer. For a conformation such as the one just above, the value of the coupling constant between H1 and H2 ($J_{1,2}$) will normally be "large" (7–8 Hz) and so indicative of a *trans*-diaxial relationship:

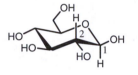

It goes almost without saying that the other, higher energy, all axial conformation will have a "small" (1–2 Hz) value for $J_{1,2}$:[c]

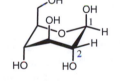

A final word of caution is necessary here — the conformation of a molecule in the solid state is not *necessarily* the same as that in the liquid or in solution.

As we saw earlier, the D-aldopentoses and D-aldohexoses exist in aqueous solution primarily as a mixture of the α- and β-pyranose forms; occasionally, as with D-ribose, -altrose, -idose and -talose, significant amounts of the furanose forms can also be found.[6] In all of these pyranose forms, it is the "normal" chair conformation that is almost always preferred; however, α- and β-D-ribose, β-D-arabinose and α-D-lyxose, -altrose and -idose all show contributions from the

---

[c] These values in carbohydrates are in general agreement with the early observations by Lemieux[4] and, a little later, with the rule enunciated by Karplus,[5] as applied to the relationship between the magnitude of the coupling constant and the size of the torsional angle between vicinal protons.

"inverted" chair conformation and, indeed, α-D-arabinose even shows a preference for it.[7]

Apart from these chair conformations for the D-aldopyranoses, there exist other, higher energy conformations, namely the **boat** and the **skew**. It must be stressed that, although these higher energy forms are not present to any extent in aqueous solution, they are discrete conformational intermediates in the conversion of one chair into the other. The **half-chair** is a common conformation for some carbohydrate derivatives where chemical modification of the pyranose ring has occurred.

What follows next is a summary of the limiting conformations for the pyranose ring, namely the chair (C), boat (B), half-chair (H) and skew (S) forms, together with their modern descriptors (it is obviously necessary to avoid such terms as "normal" and "inverted").

Chair:

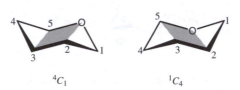

$$^4C_1 \qquad ^1C_4$$

Only two forms are possible. The descriptors arise according to the following protocol:[8]

- the lowest-numbered carbon of the ring (C1) is taken as an exoplanar atom;
- O, C2, C3, C5 define the reference plane of the chair;
- viewed clockwise (O → 2 → 3 → 5), C4 is above (below) this plane and C1 is below (above);
- atoms that are above (below) the plane are written as superscripts (subscripts) which precede (follow) the letter;
- $^4C_1$ and $^1C_4$ result.

Boat: Six forms are possible, with only two of these shown (the reference plane in each form is unique and obvious).

$$^{1,4}B \qquad B_{2,5}$$

Half-chair: Twelve forms are possible and again only two of these are shown (the reference plane is defined by four contiguous atoms and is again unique).

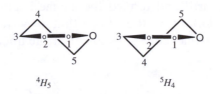

$^4H_5$ $^5H_4$

Skew: Six forms are possible, with only one of these shown (the reference plane is not obvious, being made up of three contiguous atoms and the remaining non-adjacent atom[8]).

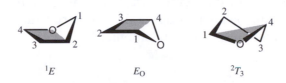

$^1S_5$

The chair form is more stable than the skew form, which is again more stable than both the boat and half-chair forms. In pyranose rings that contain a double bond, it is the half-chair that is the normal conformation.

The conformations available to the furanose ring are just the envelope (E) and the twist (T); both have ten possibilities and the energy differences among all of the conformations are quite small.

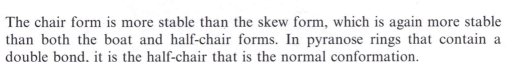

^1E E_O $^2T_3$

Let us now take time to reflect on the familiar equilibrium that is established when D-(+)-glucose is dissolved in water:

36%   acyclic   64%

The two main components of the mixture are present in the indicated amounts and each in the preferred $^4C_1$ conformation. The free energy difference for such an equilibrium amounts to about 1.5 kJ mol^{-1} in favour of the β-anomer, somewhat short of the accepted value (3.8 kJ mol^{-1}) for an equatorial over an axial hydroxyl group, the only difference between the two molecules in question. This propensity for the formation of the α-anomer over that which would normally be expected was first noted by Edward[9] and termed the **anomeric effect** by Lemieux.[10,11] So wide ranging and important is the effect that it virtually ensures the axial configuration of an electronegative substituent at the anomeric carbon:

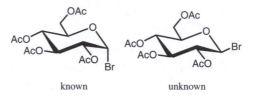

known   unknown

As well, the anomeric effect is responsible for the stabilization of conformations which would otherwise seemingly capitulate to other, unfavourable interactions:

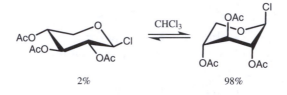

2%   98%

The origin of the anomeric effect, which itself increases with the electronegativity of the substituent and decreases in solvents of high dielectric constant, has been explained in several ways, including unfavourable lone pair–lone pair or dipole–dipole interactions in the equatorial anomer and favourable dipole–dipole interactions in the axial anomer:[12]

However, the most favoured and accepted explanation involves the interaction between a lone pair of electrons located "axially" in a molecular orbital (n) on O5 and an unoccupied, anti-bonding molecular orbital (σ^*) of the Cl–X bond.[13]

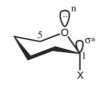

The "anti-periplanar" arrangement found in the axial anomer favours this "back-bonding", resulting in a slight shortening of the O5–Cl bond, a slight lengthening of the Cl–X bond and a general increase in the electron density at X. The causes of the anomeric effect continue to be discussed.[14–17]

The **reverse anomeric effect**[18,19] places an electropositive group at the anomeric carbon in a favoured equatorial disposition; for example, *N*-(tetra-*O*-acetyl-α-D-glucopyranosyl)-4-methylpyridinium bromide exists essentially in the normally unfavoured $^1C_4$ conformation:

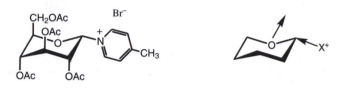

The obvious explanation for the origin of this "reverse" effect is a simple and favourable interaction of opposing dipoles. There has been a great deal of discussion recently as to whether the reverse anomeric effect even exists.[20–23]

Probably one of the greatest contributions by Lemieux to the field of carbohydrate chemistry has been his delineation of the importance of the ***exo*-anomeric effect**.[24–27] In a simple acetal derived from a carbohydrate, there operates the normal anomeric effect which stabilizes the axial anomer over the equatorial anomer.

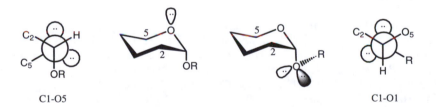

C1-O5   C1-O1

However, in the appropriate conformation of the exocyclic alkoxy group, there is again an anti-periplanar arrangement of a lone pair on oxygen (of OR) and the Cl–O5 bond — the so-called "*exo*-anomeric effect".[d] Because these two anomeric effects operate in opposite directions, the *exo*-anomeric effect is not considered important with such axial acetals. However, in an equatorial acetal, where there is no contribution from a normal anomeric effect, it is the

---

[d] As a consequence of this new term, the "anomeric effect" is sometimes referred to as the "*endo*-anomeric effect".

exo-anomeric effect that is dominant and dictates the preferred conformation of the alkoxy group at the anomeric carbon atom:

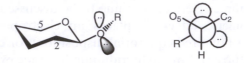

Taken to its logical conclusion, the *exo*-anomeric effect appears to be responsible for the helical shape of many polysaccharide chains — it is certainly important in determining the shape of many biologically important **oligosaccharides**.[28,e]

We have seen in these chapters that the seminal studies of Fischer were carried along in the early part of the next century by people such as Haworth and Hudson. However, it was to be Lemieux[29,f] who would dominate carbohydrate chemistry for the major part of the twentieth century, with enormous contributions to nuclear magnetic resonance (n.m.r.) spectroscopy, conformational analysis, glycobiology and synthesis. His final words on the factors that govern carbohydrate/protein binding are truly memorable.[30]

References

1. Hassel, O. and Ottar, B. (1947). *Acta Chem. Scand.*, **1**, 929.
2. Barton, D. H. R. (1950). *Experientia*, **6**, 316.
3. Barton, D. H. R. (1953). *J. Chem. Soc.*, 1027.
4. Lemieux, R. U., Kullnig, R. K., Bernstein, H. J. and Schneider, W. G. (1957). *J. Am. Chem. Soc.*, **79**, 1005.
5. Karplus, M. (1963). *J. Am. Chem. Soc.*, **85**, 2870.
6. Angyal, S. J. (1984, 1991). *Adv. Carbohydr. Chem. Biochem.*, **42**, 15; **49**, 19.
7. Collins, P. M. and Ferrier, R. J. (1995). *Monosaccharides: Their Chemistry and Their Roles in Natural Product*, John Wiley & Sons, Chichester, p. 33.
8. Schwarz, J. C. P. (1973). *J. Chem. Soc., Chem. Commun.*, 505.
9. Edward, J. T. (1955). *Chem. Ind.*, 1102.
10. Lemieux, R. U. (1964). In *Molecular Rearrangements*, de Mayo, P. ed., Interscience Publishers: John Wiley and Sons, New York, p. 709.
11. Lemieux, R. U. (1971). *Pure Appl. Chem.*, **25**, 527.
12. Wolfe, S., Rauk, A., Tel, L. M. and Csizmadia, I. G. (1971). *J. Chem. Soc. B*, 136.
13. Juaristi, E. and Cuevas, G. (1992). *Tetrahedron*, **48**, 5019.
14. Ma, B., Schaefer, H. F., III and Allinger, N. L. (1998). *J. Am. Chem. Soc.*, **120**, 3411.
15. Thatcher, G. R. J. (1993). *The Anomeric Effect and Associated Stereoelectronic Effects*, ACS Symposium Series 539, American Chemical Society, Washington DC.
16. Juaristi, E. and Cuevas, G. (1995). *The Anomeric Effect*, CRC Press, Boca Raton.

---

[e] A general term for small chains of monosaccharides, with up to ten residues in the chain.

[f] Raymond U. Lemieux (1920–2000), PhD under C. B. Purves (McGill University).

17. Box, V. G. S. (1998). *Heterocycles*, **48**, 2389.
18. Lemieux, R. U. and Morgan, A. R. (1965). *Can. J. Chem.*, **43**, 2205.
19. West, A. C. and Schuerch, C. (1973). *J. Am. Chem. Soc.*, **95**, 1333.
20. Perrin, C. L., Fabian, M. A., Brunckova, J. and Ohta, B. K. (1999). *J. Am. Chem. Soc.*, **121**, 6911.
21. Perrin, C. L. (1995). *Tetrahedron*, **51**, 11901.
22. Vaino, A. R., Chan, S. S. C., Szarek, W. A. and Thatcher, G. R. J. (1996). *J. Org. Chem.*, **61**, 4514.
23. Randell, K. D., Johnston, B. D., Green, D. F. and Pinto, B. M. (2000). *J. Org. Chem.*, **65**, 220.
24. Lemieux, R. U., Pavia, A. A., Martin, J. C. and Watanabe, K. A. (1969). *Can. J. Chem.*, **47**, 4427.
25. Praly, J.-P. and Lemieux, R. U. (1987). *Can. J. Chem.*, **65**, 213.
26. Tvaroška, I. and Bleha, T. (1989). *Adv. Carbohydr. Chem. Biochem.*, **47**, 45.
27. Tvaroska, I. and Carver, J. P. (1998). *Carbohydr. Res.*, **309**, 1.
28. Meyer, B. (1990). *Topics Curr. Chem.*, **154**, 141.
29. Lemieux, R. U. (1990). *Explorations with Sugars: How Sweet it Was*, American Chemical Society, Washington DC.
30. Lemieux, R. U. (1996). *Acc. Chem. Res.*, **29**, 373.

Chapter 6

Synthesis and Protecting Groups[1-5]

The study of carbohydrates would be a simple matter if it were confined to the natural and abundant aldoses, ketoses and oligosaccharides. However, there often arises the need for modified monosaccharides or, perhaps, an unusual or rare oligosaccharide. How would one approach the synthesis of such molecules, say, in the first instance, as "3-deoxy-D-glucose":[a]

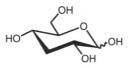

The problems are two-fold: first, the need for a chemical reaction that will replace a hydroxyl group by a hydrogen atom; second, the need to carry out this replacement *only* at C3.

Also, what of the synthesis of an oligosaccharide, say, a disaccharide:

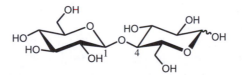

The problems are not much different from the monosaccharide example: first, a chemical method is needed to join two D-glucose units together; second, the two monosaccharides must be manipulated so that the linkage is *specifically* 1,4-β. So arise the dual needs of **synthesis**, the ability to carry out chemical reactions in carbohydrates, and **protecting groups**, those groups introduced by chemical reaction that mask one part of a molecule, yet allow access to another. The ensuing chapters will cover these two enmeshed concepts in some detail.

---

[a] As "3-deoxy-D-allose" is just as good a name, an unambiguous name should be used: 3-deoxy-D-*ribo*-hexose. The molecule is depicted as an α/β mixture of pyranose forms.

To set the stage, consider a very early synthesis, performed by Fischer in 1893:

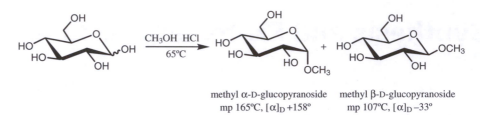

methyl α-D-glucopyranoside   methyl β-D-glucopyranoside
mp 165°C, [α]$_D$ +158°   mp 107°C, [α]$_D$ –33°

By heating D-glucose with methanol containing some hydrogen chloride, two new chemicals, actually anomeric acetals, were formed — a "synthesis" and, at the same time, a "protecting group" for the anomeric carbon. More about this unique and important reaction later.

References

1. Greene, T. W. and Wuts, P. G. M. (1991, 1999). *Protective Groups in Organic Synthesis*, John Wiley and Sons, New York.
2. Kocienski, P. J. (1994). *Protecting Groups*, Thieme, Stuttgart.
3. Jarowicki, K. and Kocienski, P. (2000). *J. Chem. Soc., Perkin Trans. 1*, 2495.
4. Hanson, J. R. (1999). *Protecting Groups in Organic Synthesis*, Sheffield Academic Press, Sheffield.
5. Grindley, T. B. (1996). Protecting groups in oligosaccharide synthesis, in *Modern Methods in Carbohydrate Synthesis*, Khan, S. H. and O'Neill, R. A. eds., Harwood Academic, Netherlands, p. 225.

Esters and Ethers

The primary role of esters and ethers introduced into carbohydrates is to protect the otherwise reactive hydroxyl groups. In addition, esters can play a dual role in precipitating useful chemical reactions at both anomeric and non-anomeric carbon atoms. Ethers, on the other hand, are inert groups found only at non-anomeric positions (otherwise, they would not be ethers but the more reactive acetals). Both protecting groups reduce the polarity of the carbohydrate and so allow for solubility in organic solvents.

Esters

Acetates: The acetylation of D-glucose was first performed in the mid-nineteenth century, helping to confirm the pentahydroxy nature of the molecule. Since then, three sets of conditions are commonly used for the transformation:

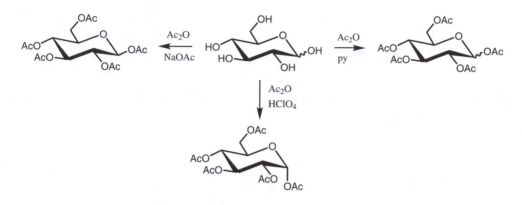

The reaction in pyridine is general and convenient and usually gives the same anomer of the penta-acetate as found in the parent free sugar.[1,2] With an acid catalyst, the reaction probably operates under thermodynamic control and gives the more stable anomer. Sodium acetate causes a rapid anomerization of the free sugar[3] and the more reactive anomer is then preferentially acetylated.[b] Iodine has recently been used for various acetylations.[6]

One of the features[c] of an O-acetyl protecting group is its ready removal to regenerate the parent alcohol — generally, the acetate is dissolved in methanol, a small piece of sodium metal is added and the required transesterification reaction is both rapid and quantitative:[7]

$$-\text{OCOCH}_3 + \text{CH}_3\text{OH} \xrightarrow{\text{CH}_3\text{ONa}} -\text{OH} + \text{CH}_3\text{COOCH}_3$$

Other systems that carry out this classical transesterification reaction are anion-exchange resin (OH⁻ form), ammonia or potassium cyanide in methanol,[2,8,9] guanidine–guanidinium nitrate in methanol[10] and a mixture of triethylamine, methanol and water.[11] For base-sensitive substrates, hydrogen chloride or tetrafluoroboric acid–ether in methanol is a viable alternative for deacetylation.[12]

For the selective acetylation of one hydroxyl group over another, one has the choice of lowering the reaction temperature or employing reagents specifically designed for such a purpose.[6,13,14] The selective removal of an acetyl group at the anomeric position can easily be achieved, probably owing to

---

[b] Deprotonation of the β-anomer of the free sugar gives a β-oxyanion which interacts unfavourably with the lone pairs of electrons on O5 — a rapid acetylation removes this interaction.[4,5]

[c] A high level of crystallinity in simple derivatives is also a much relished feature by the preparative chemist.

the better leaving group ability of the anomeric oxygen:[15–17]

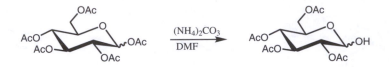

Recently, the use of enzymes, especially lipases, has added another dimension to this concept of selectivity:[18–22]

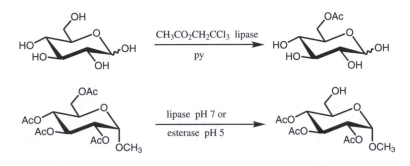

Benzoates: In general, benzoates are more robust protecting groups than acetates and often give rise to very crystalline derivatives that are useful in X-ray crystallographic determinations (for example, 4-bromobenzoates). The robustness of benzoates is reflected both in their preparation (benzoyl chloride, pyridine) and reversion to the parent alcohol (sodium-methanol for protracted periods). Acetates can be removed in preference to benzoates.[23]

The selective benzoylation of a carbohydrate[24] can be achieved either by careful control of the reaction conditions[25] or by the use of a less reactive reagent, such as N-benzoylimidazole[26,27] or 1-benzoyloxybenzotriazole:[28]

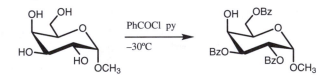

Chloroacetates: Chloroacetates are easily acquired (chloroacetic anhydride in pyridine), are stable enough to survive most synthetic transformations and can then be selectively removed (thiourea[29] or "hydrazinedithiocarbonate"[30]):

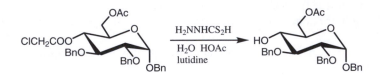

Pivaloates: Esters of pivalic acid (2,2-dimethylpropanoic acid), for the reason of steric bulk, can be installed preferentially at the more reactive sites of a sugar but require reasonably vigorous conditions for their subsequent removal:[31,32]

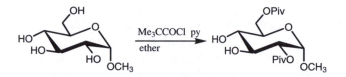

Carbonates, borates, phosphates, sulfates and nitrates: Cyclic carbonates are a sometimes-used protecting group for vicinal diols, providing the dual advantages of installation with a near neutral reagent (1,1′-carbonyldiimidazole) and removal under basic conditions.[33]

Borates, although rarely used as protecting groups, are useful in the purification, analysis and structure determination of sugar polyols. Phenylboronates seem to have more potential in synthesis.[34]

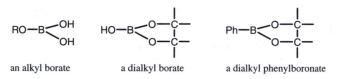

| an alkyl borate | a dialkyl borate | a dialkyl phenylboronate |

Sugar phosphates, and their oligomers, are found as the cornerstone of the molecules of life — RNA, DNA and ATP:

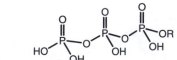

| an alkyl phosphate | a dialkyl phosphate (RNA, DNA) | an alkyl triphosphate (ATP) |

Sulfates are common components of many biologically important molecules; nitrates formed the basis of many of the early explosives.

   an alkyl sulfate   RO–NO₂   an alkyl nitrate

Sulfonates: This last group of esters is characterized not at all by its "protection" of the hydroxyl group but, rather, by its activation of the group towards nucleophilic substitution:

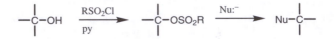

The three sulfonates commonly in question are the **tosylate** (4-toluenesulfonate), **mesylate** (methanesulfonate) and **triflate** (trifluoromethanesulfonate), generally installed in pyridine and using the acid chloride (4-toluenesulfonyl chloride and methanesulfonyl chloride) or trifluoromethanesulfonic anhydride.[35] For alcohols of low reactivity, the combination of methanesulfonyl chloride and triethylamine in dichloromethane (which produces the very reactive sulfene, CH_2SO_2) is particularly effective.[36] The sulfonates, once installed, show the following order of reactivity towards nucleophilic displacement:

$$CF_3SO_2- \gg CH_3SO_2- \geq 4\text{-}CH_3C_6H_4SO_2-$$

An addition to the above trio of sulfonates is the **imidazylate** (imidazole-sulfonate), said to be more stable than the corresponding triflate but of the same order of reactivity.[37,38]

The selective sulfonylation of a sugar polyol is possible[39] and *N*-tosylimidazole has proven to be of some use in this regard.[40]

Finally, a few general comments to end this section on esters. 4-(Dimethylamino)pyridine has proven to be an excellent adjunct in the synthesis of carbohydrate esters, especially for less reactive hydroxyl groups.[41] Acyl migration of carbohydrate esters, where possible, can be a problem but can also be put to advantage:[24,42]

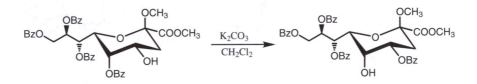

Furanosyl esters, when needed, are often best prepared indirectly from the starting sugar, for example, 1-*O*-acetyl-2,3,5-tri-*O*-benzoyl-β-D-ribose is much used in nucleoside synthesis:[43]

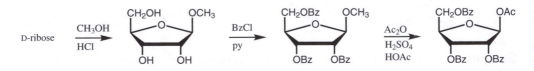

Ethers[44]

Methyl ethers: Methyl ethers are of little value as protecting groups for the hydroxyl group *per se*, as they are far too stable for easy removal, but they have a place in the history of carbohydrate chemistry in terms of structure elucidation. Since the pioneering work of Purdie (methyl iodide, silver oxide)[45] and Haworth (dimethyl sulfate, aqueous sodium hydroxide)[46] and the improvements offered by Kuhn (methyl iodide, DMF, silver oxide)[47] and Hakomori (methyl iodide, DMSO, sodium hydride),[48] "methylation analysis" has played a key role in the structure elucidation of oligosaccharides. For example, from enzyme-mediated hydrolysis studies, the naturally occurring reducing disaccharide, gentiobiose was known to consist of two β-linked D-glucose units. Complete methylation of gentiobiose gave an octamethyl "ether" which, after acid hydrolysis, yielded 2,3,4,6-tetra-*O*-methyl-D-glucose and 2,3,4-tri-*O*-methyl-D-glucose. Barring the occurrence of any outlandish ring form (a septanose), this result defined gentiobiose as 6-*O*-β-D-glucopyranosyl-D-glucopyranose:

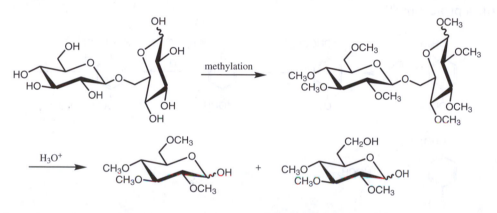

Benzyl ethers: Benzyl ethers offer a versatile means of protection for the hydroxyl group, being installed under basic (benzyl bromide, sodium hydride, DMF; benzyl bromide, sodium hydride, tetrabutylammonium iodide, THF[49,50]), acidic (benzyl trichloroacetimidate, triflic acid;[51,52] phenyldiazomethane, tetrafluoroboric acid[53]) or neutral (benzyl bromide, silver triflate) conditions.[54] As well, many methods exist for the removal of the benzyl protecting group — classical hydrogenolysis (hydrogen, palladium-on-carbon, often in the presence of an acid), catalytic transfer-hydrogenolysis (ammonium formate, palladium-on-carbon, methanol),[55,56] reduction under Birch conditions (sodium, liquid ammonia) or treatment with anhydrous ferric chloride.[57] Selective debenzylations are also possible[6,58] and trimethylsilyl triflate–acetic anhydride is a versatile reagent for the conversion of a benzyl ether into an acetate.[59]

A useful synthesis of tetra-*O*-benzyl-D-glucono-1,5-lactone is shown:

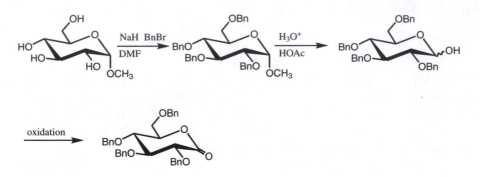

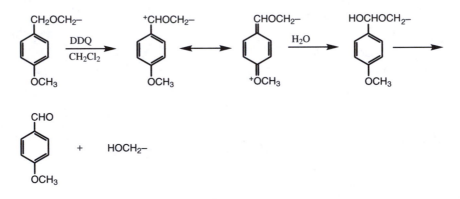

4-Methoxybenzyl ethers: This substituted benzyl ether has found an increasing use over the last two decades, for reasons of easy installation (4-methoxybenzyl chloride or bromide, sodium hydride, DMF;[60,61] 4-methoxybenzyl trichloroacetimidate[62]) and the availability of an extra, oxidative mode of deprotection:[63]

Other oxidants can also be used[60,64,65] and good selectivity is usually observed.[66] Trifluoroacetic acid and tin(IV) chloride have recently been used to remove the 4-methoxybenzyl protecting group.[67,68]

Allyl ethers:[69] Gigg, more than anyone else, has been responsible for the establishment of the allyl (prop-2-enyl) ether as a useful protecting group in carbohydrate chemistry.[70] Allyl groups may be found at both anomeric and non-anomeric positions, the latter ethers being installed under basic (allyl bromide, sodium hydride, DMF), acidic (allyl trichloroacetimidate, triflic acid)[71] or neutral conditions.[72] Many methods exist for the removal of the allyl group, most relying on an initial prop-2-enyl to prop-1-enyl isomerization[73] and varying from the classical (potassium *tert*-butoxide-dimethyl sulfoxide, followed by mercuric chloride[74] or acid[70]) to palladium- (palladium-on-carbon, acid)[75,76] and rhodium-based procedures.[77–79] Other variants of the allyl group have found some use in synthesis.[80]

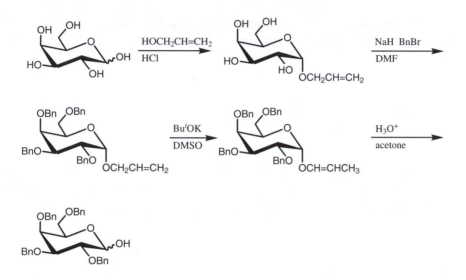

Trityl ethers: The trityl (triphenylmethyl) ether was the earliest group for the selective protection of a primary alcohol. Although the introduction of a trityl group has always been straightforward (trityl chloride, pyridine),[81] various improvements have been made.[82–84] The removal process has been much studied and the reagents used are generally either Brønsted[85] or Lewis acids;[86,87] reductive methods are occasionally used, either conventional hydrogenolysis or reduction under Birch conditions.[88]

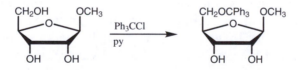

Silyl ethers:[89] The original use of silyl ethers in carbohydrates was not so much for the protection of any hydroxyl group but, rather, for the chemical modification of these normally water soluble, non-volatile compounds. For example, the *per-O*-silylation of monosaccharides was a necessary preamble to successful analysis by gas–liquid chromatography or mass spectrometry:[90]

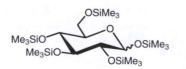

It was not until the pioneering work by Corey that silicon was used in the protection of hydroxyl groups within carbohydrates.[91] Nowadays, trimethylsilyl, triethylsilyl, *tert*-butyldimethylsilyl, *tert*-butyldiphenylsilyl and triisopropylsilyl ethers are commonly used, with normal installation *via* the chlorosilane[92,93] — quite often, the more bulky reagents show preference for a primary alcohol. Diols, especially those found in nucleosides, can be protected as a cyclic, disilyl derivative.

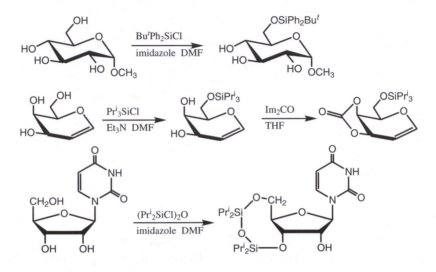

Silyl ethers survive many of the common synthetic transformations of organic chemistry[94] but are readily removed, when required, by treatment with a reagent which supplies the fluoride ion, e.g. tetrabutylammonium fluoride, hydrogen fluoride-pyridine (the Si-F bond is extremely strong, 590 kJ mol^{-1}).[89] Strongly basic conditions will cleave a silyl ether and, not surprisingly, migration of the silicon protecting group or other vulnerable residues, e.g. esters, will occur under these conditions.[95] Silyl ethers can be cleaved under acidic conditions and the general ease of acid hydrolysis is Me$_3$SiO- > Et$_3$SiO- >> Bu^{t}Me$_2$SiO- >> Pr$^i_3$SiO- >> Bu^{t}Ph$_2$SiO-. Some very mild procedures for the removal of silyl ethers have recently been reported.[96,97]

References

1. Wolfrom, M. L. and Thompson, A. (1963). *Methods Carbohydr. Chem.*, **2**, 211.
2. Conchie, J., Levvy, G. A. and Marsh, C. A. (1957). *Adv. Carbohydr. Chem.*, **12**, 157.
3. Swain, C. G. and Brown, J. F., Jr (1952). *J. Am. Chem. Soc.*, **74**, 2538.
4. Schmidt, R. R. (1986). *Angew. Chem. Int. Ed. Engl.*, **25**, 212.
5. Schmidt, R. R. and Michel, J. (1984). *Tetrahedron Lett.*, **25**, 821.
6. Kartha, K. P. R. and Field, R. A. (1997). *Tetrahedron*, **53**, 11753.

7. Zemplén, G. and Pacsu, E. (1929). *Ber. Dtsch. Chem. Ges.*, **62**, 1613.
8. Lemieux, R. U. and Stick, R. V. (1975). *Aust. J. Chem.*, **28**, 1799.
9. Herzig, J., Nudelman, A., Gottlieb, H. E. and Fischer, B. (1986). *J. Org. Chem.*, **51**, 727.
10. Ellervik, U. and Magnusson, G. (1997). *Tetrahedron Lett.*, **38**, 1627.
11. Lemieux, R. U., Hendriks, K. B., Stick, R. V. and James, K. (1975). *J. Am. Chem. Soc.*, **97**, 4056.
12. Vekemans, J. A. J. M., Franken, G. A. M., Chittenden, G. J. F. and Godefroi, E. F. (1987). *Tetrahedron Lett.*, **28**, 2299.
13. Sinaÿ, P., Jacquinet, J.-C., Petitou, M., Duchaussoy, P., Lederman, I., Choay, J. and Torri, G. (1984). *Carbohydr. Res.*, **132**, C5.
14. Ishihara, K., Kurihara, H. and Yamamoto, H. (1993). *J. Org. Chem.*, **58**, 3791.
15. Best, W. M., Dunlop, R. W., Stick, R. V. and White, S. T. (1994). *Aust. J. Chem.*, **47**, 433.
16. Mikamo, M. (1989). *Carbohydr. Res.*, **191**, 150.
17. Grynkiewicz, G., Fokt, I., Szeja, W. and Fitak, H. (1989). *J. Chem. Res., Synop.*, 152.
18. Gridley, J. J., Hacking, A. J., Osborn, H. M. I. and Spackman, D. G. (1998). *Tetrahedron*, **54**, 14925.
19. Therisod, M. and Klibanov, A. M. (1986). *J. Am. Chem. Soc.*, **108**, 5638.
20. Horrobin, T., Tran, C. H. and Crout, D. (1998). *J. Chem. Soc., Perkin Trans. 1*, 1069.
21. Bashir, N. B., Phythian, S. J., Reason, A. J. and Roberts, S. M. (1995). *J. Chem. Soc., Perkin Trans. 1*, 2203.
22. Hennen, W. J., Sweers, H. M., Wang, Y.-F. and Wong, C.-H. (1988). *J. Org. Chem.*, **53**, 4939.
23. Byramova, N. E., Ovchinnikov, M. V., Backinowsky, L. V. and Kochetkov, N. K. (1983). *Carbohydr. Res.*, **124**, C8.
24. Haines, A. H. (1976). *Adv. Carbohydr. Chem. Biochem.*, **33**, 11.
25. Williams, J. M. and Richardson, A. C. (1967). *Tetrahedron*, **23**, 1369.
26. Carey, F. A. and Hodgson, K. O. (1970). *Carbohydr. Res.*, **12**, 463.
27. Chittenden, G. J. F. (1971). *Carbohydr. Res.*, **16**, 495.
28. Pelyvás, I. F., Lindhorst, T. K., Streicher, H. and Thiem, J. (1991). *Synthesis*, 1015.
29. Glaudemans, C. P. J. and Bertolini, M. J. (1980). *Methods Carbohydr. Chem.*, **8**, 271.
30. van Boeckel, C. A. A. and Beetz, T. (1983). *Tetrahedron Lett.*, **24**, 3775.
31. Jiang, L. and Chan, T.-H. (1998). *J. Org. Chem.*, **63**, 6035.
32. Greene, T. W. and Wuts, P. G. M. (1991). *Protective Groups in Organic Synthesis*, John Wiley & Sons, New York, p. 99.
33. Kutney, J. P. and Ratcliffe, A. H. (1975). *Synth. Commun.*, **5**, 47.
34. Cross, G. G. and Whitfield, D. M. (1998). *Synlett*, 487.
35. Binkley, R. W. and Ambrose, M. G. (1984). *J. Carbohydr. Chem.*, **3**, 1.
36. King, J. F. (1975). *Acc. Chem. Res.*, **8**, 10.
37. Hanessian, S. and Vatèle, J.-M. (1981). *Tetrahedron Lett.*, **22**, 3579.
38. Vatèle, J.-M. and Hanessian, S. (1997). Nucleophilic displacement reactions of imidazole-1-sulfonate esters, in *Preparative Carbohydrate Chemistry*, Hanessian, S. ed., Marcel Dekker, New York, p. 127.
39. Cramer, F. D. (1963). *Methods Carbohydr. Chem.*, **2**, 244.
40. Hicks, D. R. and Fraser-Reid, B. (1974). *Synthesis*, 203.
41. Höfle, G., Steglich, W. and Vorbrüggen, H. (1978). *Angew. Chem. Int. Ed. Eng.*, **17**, 569.
42. Danishefsky, S. J., DeNinno, M. P. and Chen, S.-h. (1988). *J. Am. Chem. Soc.*, **110**, 3929.
43. Recondo, E. F. and Rinderknecht, H. (1959). *Helv. Chim. Acta*, **42**, 1171.
44. Staněk, J., Jr. (1990). *Top. Curr. Chem.*, **154**, 209.
45. Purdie, T. and Irvine, J. C. (1903). *J. Chem. Soc. (Trans.)*, **83**, 1021.

46. Haworth, W. N. (1915). *J. Chem. Soc. (Trans.)*, **107**, 8.
47. Kuhn, R., Baer, H. H. and Seeliger, A. (1958). *Liebigs Ann. Chem.*, **611**, 236.
48. Hakomori, S. (1964). *J. Biochem. (Tokyo)*, **55**, 205.
49. Czernecki, S., Georgoulis, C., Provelenghiou, C. and Fusey, G. (1976). *Tetrahedron Lett.*, 3535.
50. Rana, S. S., Vig, R. and Matta, K. L. (1982–83). *J. Carbohydr. Chem.*, **1**, 261.
51. Wessel, H.-P., Iversen, T. and Bundle, D. R. (1985). *J. Chem. Soc., Perkin Trans. 1*, 2247.
52. Jensen, H. S., Limberg, G. and Pedersen, C. (1997). *Carbohydr. Res.*, **302**, 109.
53. Liotta, L. J. and Ganem, B. (1989). *Tetrahedron Lett.*, **30**, 4759.
54. Berry, J. M. and Hall, L. D. (1976). *Carbohydr. Res.*, **47**, 307.
55. Anwer, M. K. and Spatola, A. F. (1980). *Synthesis*, 929.
56. Bieg, T. and Szeja, W. (1985). *Synthesis*, 76.
57. Rodebaugh, R., Debenham, J. S. and Fraser-Reid, B. (1996). *Tetrahedron Lett.*, **37**, 5477.
58. Yang, G., Ding, X. and Kong, F. (1997). *Tetrahedron Lett.*, **38**, 6725.
59. Alzeer, J. and Vasella, A. (1995). *Helv. Chim. Acta*, **78**, 177.
60. Takaku, H., Kamaike, K. and Tsuchiya, H. (1984). *J. Org. Chem.*, **49**, 51.
61. Kunz, H. and Unverzagt, C. (1992). *J. prakt. Chem.*, **334**, 579.
62. Nakajima, N., Horita, K., Abe, R. and Yonemitsu, O. (1988). *Tetrahedron Lett.*, **29**, 4139.
63. Oikawa, Y., Yoshioka, T. and Yonemitsu, O. (1982). *Tetrahedron Lett.*, **23**, 885.
64. Classon, B., Garegg, P. J. and Samuelsson, B. (1984). *Acta Chem. Scand.*, **B38**, 419.
65. Johansson, R. and Samuelsson, B. (1984). *J. Chem. Soc., Perkin Trans. 1*, 2371.
66. Horita, K., Yoshioka, T., Tanaka, T., Oikawa, Y. and Yonemitsu, O. (1986). *Tetrahedron*, **42**, 3021.
67. Yan, L. and Kahne, D. (1995). *Synlett*, 523.
68. Yu, W., Su, M., Gao, X., Yang, Z. and Jin, Z. (2000). *Tetrahedron Lett.*, **41**, 4015.
69. Guibé, F. (1997). *Tetrahedron*, **53**, 13509.
70. Gigg, J. and Gigg, R. (1966). *J. Chem. Soc. C*, 82.
71. Wessel, H.-P. and Bundle, D. R. (1985). *J. Chem. Soc., Perkin Trans. 1*, 2251.
72. Lakhmiri, R., Lhoste, P. and Sinou, D. (1989). *Tetrahedron Lett.*, **30**, 4669.
73. Gent, P. and Gigg, R. (1974). *J. Chem. Soc., Chem. Commun.*, 277.
74. Gigg, R. and Warren, C. D. (1968). *J. Chem. Soc. C*, 1903.
75. Boss, R. and Scheffold, R. (1976). *Angew. Chem. Int. Ed. Engl.*, **15**, 558.
76. Nukada, T., Kitajima, T., Nakahara, Y. and Ogawa, T. (1992). *Carbohydr. Res.*, **228**, 157.
77. Corey, E. J. and Suggs, J. W. (1973). *J. Org. Chem.*, **38**, 3224.
78. Ziegler, F. E., Brown, E. G. and Sobolov, S. B. (1990). *J. Org. Chem.*, **55**, 3691.
79. Boons, G.-J., Burton, A. and Isles, S. (1996). *Chem. Commun.*, 141.
80. Gigg, R. (1980). *J. Chem. Soc., Perkin Trans. 1*, 738.
81. Helferich, B. (1948). *Adv. Carbohydr. Chem.*, **3**, 79.
82. Hanessian, S. and Staub, A. P. A. (1976). *Methods Carbohydr. Chem.*, **7**, 63.
83. Chaudhary, S. K. and Hernandez, O. (1979). *Tetrahedron Lett.*, 95.
84. Murata, S. and Noyori, R. (1981). *Tetrahedron Lett.*, **22**, 2107.
85. Krainer, E., Naider, F. and Becker, J. (1993). *Tetrahedron Lett.*, **34**, 1713.
86. Kohli, V., Blöcker, H. and Köster, H. (1980). *Tetrahedron Lett.*, **21**, 2683.
87. Randazzo, G., Capasso, R., Cicala, M. R. and Evidente, A. (1980). *Carbohydr. Res.*, **85**, 298.
88. Kováč, P. and Bauer, S. (1972). *Tetrahedron Lett.*, 2349.
89. Greene, T. W. and Wuts, P. G. M. (1991). *Protective Groups in Organic Synthesis*, John Wiley & Sons, New York, p. 68.
90. Dutton, G. G. S. (1973). *Adv. Carbohydr. Chem. Biochem.*, **28**, 11.
91. Corey, E. J. and Venkateswarlu, A. (1972). *J. Am. Chem. Soc.*, **94**, 6190.

92. Lalonde, M. and Chan, T. H. (1985). *Synthesis*, 817.
93. Danishefsky, S. J. and Bilodeau, M. T. (1996). *Angew. Chem. Int. Ed. Engl.*, **35**, 1380.
94. Muzart, J. (1993). *Synthesis*, 11.
95. Mulzer, J. and Schöllhorn, B. (1990). *Angew. Chem. Int. Ed. Engl.*, **29**, 431.
96. Feixas, J., Capdevila, A., Camps, F. and Guerrero, A. (1992). *J. Chem. Soc., Chem. Commun.*, 1451.
97. Farràs, J., Serra, C. and Vilarrasa, J. (1998). *Tetrahedron Lett.*, **39**, 327.

Acetals[1–6]

Before embarking on a discussion of carbohydrate acetals, it is timely to review the reactivity of the various hydroxyl groups within D-glucopyranose:

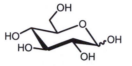

Of the five hydroxyl groups present, it is the *anomeric* hydroxyl group that is unique, being part of a *hemiacetal* structure — all of the other hydroxyl groups show the reactions typical of an alcohol. We have already seen several unique reactions of the anomeric centre, one of which was the formation (by Fischer) of a mixture of *acetals* by the treatment of D-glucose with methanol and hydrogen chloride:

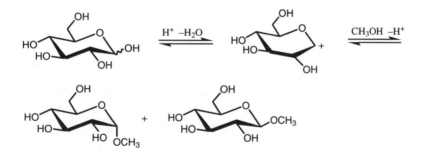

These methyl acetals, methyl α- and β-D-glucopyranoside, offered a form of protection to the anomeric centre and allowed for the useful synthesis of protected, free sugars:

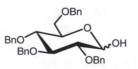

Other acetals have been developed which also offer this unique protection of the anomeric centre but have the added advantage of removal under milder and more selective conditions:

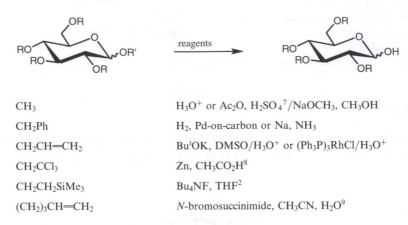

R' is	CH$_3$	H$_3$O$^+$ or Ac$_2$O, H$_2$SO$_4$[7]/NaOCH$_3$, CH$_3$OH
	CH$_2$Ph	H$_2$, Pd-on-carbon or Na, NH$_3$
	CH$_2$CH=CH$_2$	Bu^{t}OK, DMSO/H$_3$O$^+$ or (Ph$_3$P)$_3$RhCl/H$_3$O$^+$
	CH$_2$CCl$_3$	Zn, CH$_3$CO$_2$H[8]
	CH$_2$CH$_2$SiMe$_3$	Bu$_4$NF, THF[2]
	(CH$_2$)$_3$CH=CH$_2$	N-bromosuccinimide, CH$_3$CN, H$_2$O[9]

Acetals, apart from being useful in the protection of the anomeric centre, may, in a "peripheral" sense, be used for the protection of other hydroxyl groups:

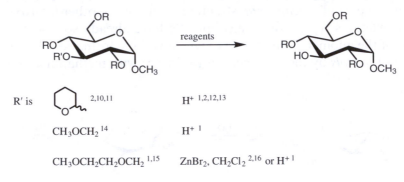

R' is		
(pyranyl) [2,10,11]		H$^+$ [1,2,12,13]
CH$_3$OCH$_2$ [14]		H$^+$ [1]
CH$_3$OCH$_2$CH$_2$OCH$_2$ [1,15]	ZnBr$_2$, CH$_2$Cl$_2$ [2,16] or H$^+$ [1]

Even though these sorts of acetals find great use in general synthetic chemistry, their use and acceptance has been somewhat limited in carbohydrates — perhaps the reasons for this can be found in the pages which follow.

Cyclic Acetals

Any synthetic endeavour with carbohydrates must recognize the presence, more often than not, of molecules containing more than one hydroxyl group, often in *cis*-1,2- or 1,3-dispositions. So arose the need to "protect" such diol systems and "cyclic acetals" were the obvious answer. The **benzylidene** and **isopropylidene** acetal groups stand (almost) alone as two prodigious protecting groups of diols and some general comments are warranted.

In line with the general principles of stereochemistry and conformational analysis,[17] the cyclic acetals of benzaldehyde (benzylidene) and acetone (isopropylidene), when formed under equilibrating conditions, generally result where possible in 1,3-dioxane and 1,3-dioxolane structures, respectively:

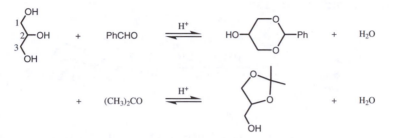

In addition, under these equilibrating conditions, the phenyl group will strive to take up an equatorial positioning:

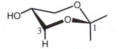

However, situations sometimes arise where it is necessary to protect a 1,3-diol as an isopropylidene acetal — then, a reagent must be found which will provide the acetal under non-equilibrating conditions, bearing in mind that the product will suffer from destabilizing "1,3-diaxial" interactions:

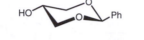

Benzylidene acetals: The treatment of a carbohydrate diol with benzaldehyde under a variety of acidic conditions,[18,19] typically utilizing fused zinc chloride, furnishes the benzylidene acetal in excellent yield:

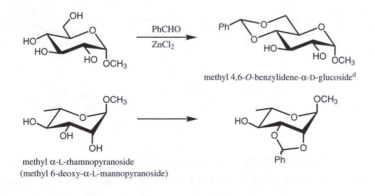

methyl 4,6-O-benzylidene-α-D-glucoside[d]

methyl α-L-rhamnopyranoside
(methyl 6-deoxy-α-L-mannopyranoside)

---

[d] Note that, in the name, the configuration (R) of the new acetal centre is not specified but presumed, and the use of "glucopyranoside" is an unnecessary tautology.

When this old but reliable method fails, one may resort to a "transacetalization" process involving the treatment of the diol with benzaldehyde dimethyl acetal under acidic conditions:[20-22]

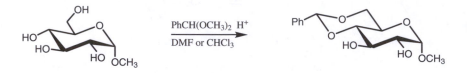

Finally, when a benzylidene acetal needs to be installed under non-acidic conditions, α,α-dibromotoluene in pyridine can be used; this is not a common method as, not surprisingly, a mixture of diastereoisomers often results:[23]

One of the strengths of the benzylidene acetal protecting group is that it may be removed by the normal reagents (acid treatment,[24,25] hydrogenolysis or reduction under Birch conditions)[1-5] to regenerate the parent diol or, more productively, by methods which involve functional group transformations. Over the past two decades, an array of methods has been devised for the removal of the benzylidene acetal group with concomitant conversion into a benzyl ether, for example:[26,27]

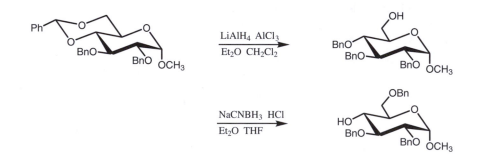

The methods are based on preferential complexation (at O6) or protonation (at O4), leading to intermediate carbocations that are subsequently reduced:

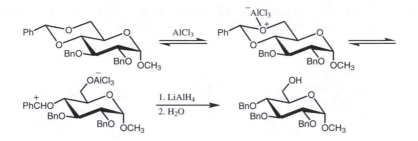

A summary of the methods currently in use is shown in Table 1.

Table 1

Electrophile	Reducing agent	Solvent	Product	Reference
AlCl$_3$	LiAlH$_4$	Et$_2$O, CH$_2$Cl$_2$	6-OH	28
Ph$_2$BBr	PhSH or THF.BH$_3$	CH$_2$Cl$_2$	6-OH	29
Bu$_2$BOTf	THF.BH$_3$	CH$_2$Cl$_2$	6-OH	30
AlCl$_3$	Me$_3$NBH$_3$	PhCH$_3$ or CH$_2$Cl$_2$	6-OH	31
		THF	4-OH	
Et$_2$OBF$_3$	Me$_2$NHBH$_3$	CH$_2$Cl$_2$	6-OH	32
		CH$_3$CN	4-OH	
HCl	NaCNBH$_3$	THF	4-OH	33
CF$_3$COOH	Et$_3$SiH	CH$_2$Cl$_2$	4-OH	34
Et$_2$OBF$_3$				35
CF$_3$SO$_3$H	NaCNBH$_3$	THF	4-OH	36

No real mechanistic studies have been performed on the reductive opening of benzylidene acetals but it is obvious that the process is governed by a complex interplay among steric, acid-base and solvent effects.[26,31] Finally, the reaction is not restricted just to dioxane-type benzylidene acetals — some very interesting observations have been made with dioxolane acetals:[28]

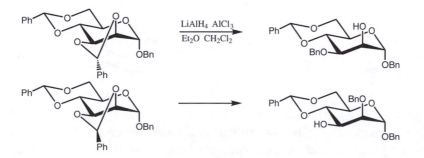

Another useful transformation of benzylidene acetals involves treatment with *N*-bromosuccinimide, to form a bromo benzoate:[37–40]

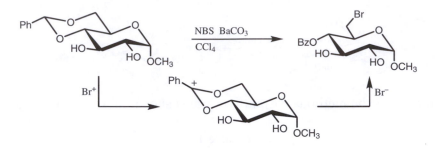

The use of calcium carbonate instead of barium carbonate seems to improve the process[41] and a related photochemical version employing bromotrichloromethane has been reported.[42]

Finally, another oxidative method employs ozone to convert a benzylidene acetal into a hydroxy benzoate:[43]

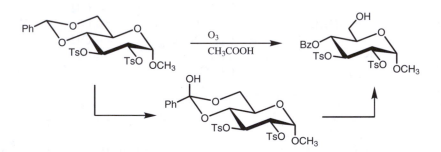

4-Methoxybenzylidene acetals: The 4-methoxybenzylidene acetal is usually prepared from the carbohydrate diol and 4-methoxybenzaldehyde dimethyl acetal under acidic conditions:[44]

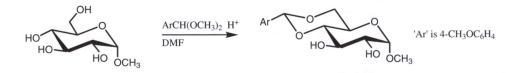

An advantage possessed by this substituted benzylidene acetal, apart from the increased lability to acid, is the somewhat milder conditions for reductive ring-opening:[44]

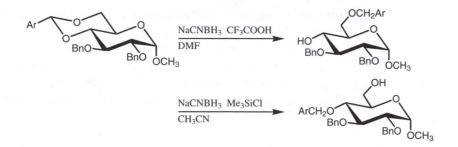

Isopropylidene acetals: The first isopropylidene acetal of a sugar was prepared by Fischer in 1895[45] and, since then, three main methods have emerged for the installation of this important protecting group under acidic conditions, utilizing acetone, 2,2-dimethoxypropane or 2-methoxypropene.

Nothing warms the heart of a carbohydrate chemist more than the sight of the following three classical transformations:[46]

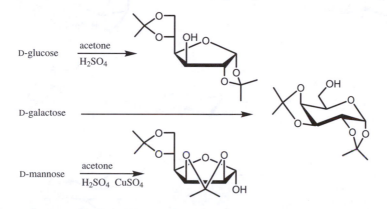

Under the strongly acidic conditions employed (sulfuric acid), all three products are the thermodynamically favoured ones and, in one step, provide direct access to molecules with just one hydroxyl group available for subsequent transformations. Other protic (HBF$_4$-ether,[21] 4-toluenesulfonic acid[47]) and some Lewis (FeCl$_3$[48]) acids promote the acetalization process equally well.

2,2-Dimethoxypropane generally gives similar results to those with acetone but useful differences are often observed:[49]

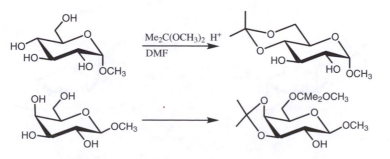

The latter transformation gives a rapid, direct and high yielding route into a D-galactose unit suitable for further elaboration just at O2.

The last reagent, 2-methoxypropene, was developed in the mid-1970s, largely by the efforts of Gelas and Horton, for the synthesis of isopropylidene acetals. However, owing to the high reactivity of the reagent and the trace amounts of acid catalyst used, the products formed were those ascribed to "kinetic control":[50]

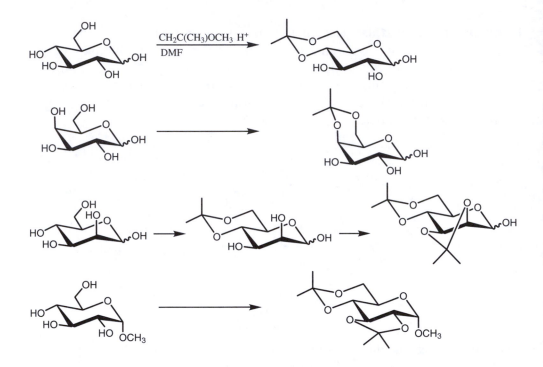

Finally, the removal of the isopropylidene protecting group generally offers few problems — aqueous acid (trifluoroacetic acid–water, 9:1 is particularly effective) is commonly used, being selective for some di-O-isopropylidene derivatives; other occasions may warrant the use of iodine in methanol[51] or a Lewis acid such as iron(III) chloride[52] or copper(II) chloride.[53]

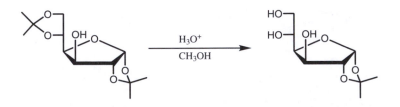

Diacetals: One of the triumphs of modern carbohydrate chemistry has been to attract "into the fold", as it were, outstanding synthetic chemists from mainstream organic chemistry. A major reason for this attraction has been the occurrence of carbohydrates in various natural products and the role that carbohydrates play in many biological processes. These gifted chemists have been able to view carbohydrates in an "unbiased" light and so make advances in areas that may have appeared somewhat stagnant.

In the area of acetal protecting groups, Ley has published an elegant sequence of papers, which describes new methods for the protection of diequatorial vicinal diols, as commonly found in carbohydrates. In the early publications, a bisdihydropyran reagent was able to react with just the 2,3-diol of methyl α-D-galactopyranoside, by virtue of forming a dispiroacetal that is uniquely stabilized by *four* individual anomeric effects, a *trans*-decalin core and four equatorial substituents on the central dioxane ring:[54]

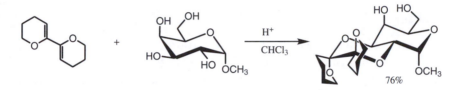

Some limitations were observed with the reaction of various alkyl α-D-mannopyranosides and the bisdihydropyran reagent and, in general, quite acidic conditions were needed to remove the dispiroacetal protecting group.[55]

In an improvement to the whole procedure, it was found that 1,1,2,2-tetramethoxycyclohexane offered the same selectivity for diequatorial vicinal diols, including those of methyl α-D-mannopyranoside:[56]

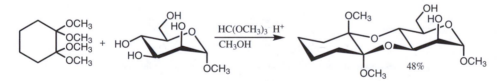

Finally, the reagent of choice for the protection of a diequatorial vicinal diol was found to be not a diacetal at all but, rather, a diketone:[57]

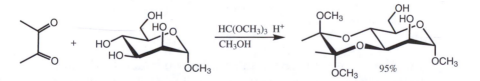

This most remarkable reaction is destined to become of great use in synthetic carbohydrate chemistry.

Cyclohexylidene acetals: The cyclohexylidene acetal is a sometimes used protecting group (partly because the resulting n.m.r. spectra are quite complex) that offers ease of installation (cyclohexanone, cyclohexanone dimethyl acetal or 1-methoxycyclohexene under acidic conditions), a propensity to form 1,3-dioxolanes where possible and a greater stability towards hydrolysis than the corresponding isopropylidene acetal:[58]

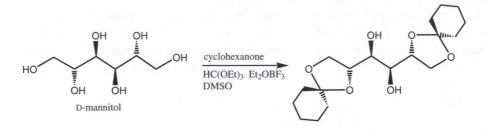

Dithioacetals[59]

Anomeric dithioacetals, since their first preparation by Fischer in 1894,[60] have maintained their importance to synthetic chemists because they offer one of the few ways of locking an aldose in its acyclic form.[61] Subsequent manipulations on the rest of the molecule can offer useful synthetic intermediates:[62]

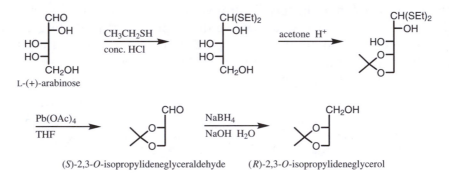

(S)-2,3-O-isopropylideneglyceraldehyde   (R)-2,3-O-isopropylideneglycerol

It is an unfortunate fact that removal of the dithioacetal protecting group, when necessary, often requires the use of environmentally unfriendly heavy metal salts, such as Hg(II). Hence, other methods have been devised.[63]

Thioacetals

Although there has been some recent interest shown in acyclic thioacetals,[64] it is the cyclic thioacetals, or 1-thio sugars, that are the most important member of this class. As such, these thioacetals are versatile starting materials for the synthesis of disaccharides and higher oligomers and owe their popularity to the ease of preparation and handling:[65,66]

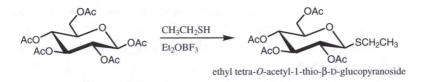

ethyl tetra-*O*-acetyl-1-thio-β-D-glucopyranoside

Stannylene Acetals[67–69]

The treatment of a vicinal diol with dibutyltin oxide gives rise to a cyclic derivative known as a "stannylene acetal":

Apparently, the size of the tin atom allows such stannylene acetals to form from both *cis* and *trans* vicinal diols; as well, the tin atom causes an increase in the reactivity (nucleophilicity) of an attached oxygen atom so that subsequent acylations and alkylations may be performed under very mild conditions:

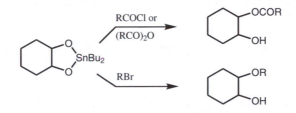

Not surprisingly, this sequence of reactions has found great application in the selective protection of carbohydrate diols and polyols:[70]

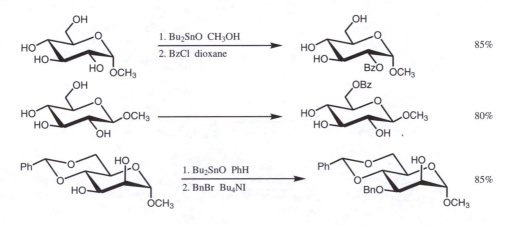

The above transformations show that, even though the acylation/alkylation is regioselective, it is not always possible to predict the outcome of a particular reaction. In general, an equatorial oxygen is functionalized in preference to one that is axial[71] and the necessary addition of a tetrabutylammonium halide increases the rate of the alkylation reaction.[72,73] In addition, 1,3-diol systems seem able to form a cyclic, stannylene acetal.

Two recent and conflicting publications, both employing dibutyltin dimethoxide as the reagent, have highlighted the care that must be taken in making generalizations about this particularly useful synthetic method.[74,75] A regioselective sulfation of disaccharides that uses stannylene acetal methodology has been reported.[76,77] Finally, a report on anomeric stannylene acetals allows for the isomerization of 6-O-trityl-D-galactose into the rare sugar, D-talose:[78]

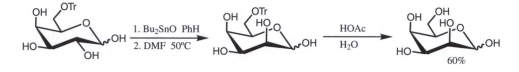

Shortly after the establishment of the stannylene acetal methodology, it was found that the treatment of an alcohol with bis(tributyltin) oxide gave rise to a "stannyl ether".[79]

$$(Bu_3Sn)_2O + 2\ HOR \rightarrow 2\ Bu_3SnOR + H_2O$$

Again, the reactivity of the oxygen in the stannyl ether was greatly enhanced, sufficiently so as to be able to react directly with acylating agents but again needing the presence of a tetrabutylammonium halide for successful alkylation.[80] Some interesting transformations of carbohydrate polyols were observed:

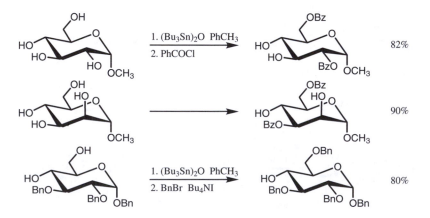

A recent comment has been made on the variability of the regioselectivity of the process according to the reaction conditions employed.[81]

References

1. Greene, T. W. and Wuts, P. G. M. (1991, 1999). *Protective Groups in Organic Synthesis*, John Wiley & Sons, New York.
2. Kocienski, P. J. (1994). *Protecting Groups*, Thieme, Stuttgart, pp. 68, 96.
3. Gelas, J. (1981). *Adv. Carbohydr. Chem. Biochem.*, **39**, 71.
4. Calinaud, P. and Gelas, J. (1997). Synthesis of isopropylidene, benzylidene, and related acetals; in *Preparative Carbohydrate Chemistry*, Hanessian, S. ed., Marcel Dekker, New York, p. 3.
5. Haines, A. H. (1981). *Adv. Carbohydr. Chem. Biochem.*, **39**, 13.
6. Staněk, J., Jr. (1990). *Top. Curr. Chem.*, **154**, 209.
7. Guthrie, R. D. and McCarthy, J. F. (1967). *Adv. Carbohydr. Chem.*, **22**, 11.
8. Lemieux, R. U. and Driguez, H. (1975). *J. Am. Chem. Soc.*, **97**, 4069.
9. Fraser-Reid, B., Udodong, U. E., Wu, Z., Ottosson, H., Merritt, J. R., Rao, C. S., Roberts, C. and Madsen, R. (1992). *Synlett*, 927.
10. Wolfrom, M. L., Beattie, A., Bhattacharjee, S. S. and Parekh, G. G. (1968). *J. Org. Chem.*, **33**, 3990.
11. Miyashita, N., Yoshikoshi, A. and Grieco, P. A. (1977). *J. Org. Chem.*, **42**, 3772.
12. Oriyama, T., Yatabe, K., Sugawara, S., Machiguchi, Y. and Koga, G. (1996). *Synlett*, 523.
13. Zhang, Z.-H., Li, T.-S., Jin, T.-S. and Wang, J.-X. (1998). *J. Chem. Res., Synop.*, 152.
14. Nishino, S. and Ishido, Y. (1986). *J. Carbohydr. Chem.*, **5**, 313.
15. El-Shenawy, H. A. and Schuerch, C. (1984). *Carbohydr. Res.*, **131**, 239.
16. Guindon, Y., Morton, H. E. and Yoakim, C. (1983). *Tetrahedron Lett.*, **24**, 3969.
17. Eliel, E. L., Wilen, S. H. and Mander, L. N. (1994). *Stereochemistry of Organic Compounds*, John Wiley & Sons, New York.
18. Richtmyer, N. K. (1962). *Methods Carbohydr. Chem.*, **1**, 107.
19. Chittenden, G. J. F. (1988). *Rec. Trav. Chim. Pays-Bas*, **107**, 607.
20. Evans, M. E. (1980). *Methods Carbohydr. Chem.*, **8**, 313.
21. Albert, R., Dax, K., Pleschko, R. and Stütz, A. E. (1985). *Carbohydr. Res.*, **137**, 282.
22. Ferro, V., Mocerino, M., Stick, R. V. and Tilbrook, D. M. G. (1988). *Aust. J. Chem.*, **41**, 813.
23. Garegg, P. J. and Swahn, C.-G. (1980). *Methods Carbohydr. Chem.*, **8**, 317.
24. Xia, J. and Hui, Y. (1996). *Synth. Commun.*, **26**, 881.
25. Andrews, M. A. and Gould, G. L. (1992). *Carbohydr. Res.*, **229**, 141.
26. Garegg, P. J. (1997). Regioselective cleavage of *O*-benzylidene acetals to benzyl ethers, in *Preparative Carbohydrate Chemistry*, Hanessian, S. ed., Marcel Dekker, New York, p. 53.
27. Garegg, P. J. (1984). *Pure Appl. Chem.*, **56**, 845.
28. Lipták, A., Imre, J., Harangi, J., Nánási, P. and Neszmélyi, A. (1982). *Tetrahedron*, **38**, 3721.
29. Guindon, Y., Girard, Y., Berthiaume, S., Gorys, V., Lemieux, R. and Yoakim, C. (1990). *Can. J. Chem.*, **68**, 897.
30. Jiang, L. and Chan, T.-H. (1998). *Tetrahedron Lett.*, **39**, 355.
31. Ek, M., Garegg, P. J., Hultberg, H. and Oscarson, S. (1983). *J. Carbohydr. Chem.*, **2**, 305.
32. Oikawa, M., Liu, W.-C., Nakai, Y., Koshida, S., Fukase, K. and Kusumoto, S. (1996). *Synlett*, 1179.
33. Garegg, P. J., Hultberg, H. and Wallin, S. (1982). *Carbohydr. Res.*, **108**, 97.
34. DeNinno, M. P., Etienne, J. B. and Duplantier, K. C. (1995). *Tetrahedron Lett.*, **36**, 669.

35. Debenham, S. D. and Toone, E. J. (2000). *Tetrahedron:* Asymmetry, **11**, 385.
36. Pohl, N. L. and Kiessling, L. L. (1997). *Tetrahedron Lett.*, **38**, 6985.
37. Hanessian, S. and Plessas, N. R. (1969). *J. Org. Chem.*, **34**, 1053.
38. Hullar, T. L. and Siskin, S. B. (1970). *J. Org. Chem.*, **35**, 225.
39. Hanessian, S. (1972). *Methods Carbohydr. Chem.*, **6**, 183.
40. Hanessian, S. (1987). *Org. Synth.*, **65**, 243.
41. Chrétien, F., Khaldi, M. and Chapleur, Y. (1990). *Synth. Commun.*, **20**, 1589.
42. Chana, J. S., Collins, P. M., Farnia, F. and Peacock, D. J. (1988). *J. Chem. Soc., Chem. Commun.*, 94.
43. Deslongchamps, P., Moreau, C., Fréhel, D. and Chênevert, R. (1975). *Can. J. Chem.*, **53**, 1204.
44. Johansson, R. and Samuelsson, B. (1984). *J. Chem. Soc., Perkin Trans. 1*, 2371.
45. Fischer, E. (1895). *Ber. Dtsch. Chem. Ges.*, **28**, 1145.
46. Schmidt, O. Th. (1963). *Methods Carbohydr. Chem.*, **2**, 318.
47. He, D. Y., Li, Z. J., Li, Z. J., Liu, Y. Q., Qiu, D. X. and Cai, M. S. (1992). *Synth. Commun.*, **22**, 2653.
48. Singh, P. P., Gharia, M. M., Dasgupta, F. and Srivastava, H. C. (1977). *Tetrahedron Lett.*, 439.
49. Catelani, G., Colonna, F. and Marra, A. (1988). *Carbohydr. Res.*, **182**, 297.
50. Gelas, J. and Horton, D. (1981). *Heterocycles*, **16**, 1587.
51. Szarek, W. A., Zamojski, A., Tiwari, K. N. and Ison, E. R. (1986). *Tetrahedron Lett.*, **27**, 3827.
52. Kim, K. S., Song, Y. H., Lee, B. H. and Hahn, C. S. (1986). *J. Org. Chem.*, **51**, 404.
53. Saravanan, P., Chandrasekhar, M., Anand, R. V. and Singh, V. K. (1998). *Tetrahedron Lett.*, **39**, 3091.
54. Ley, S. V., Leslie, R., Tiffin, P. D. and Woods, M. (1992). *Tetrahedron Lett.*, **33**, 4767.
55. Ley, S. V., Boons, G.-J., Leslie, R., Woods, M. and Hollinshead, D. M. (1993). *Synthesis*, 689.
56. Grice, P., Ley, S. V., Pietruszka, J., Priepke, H. W. M. and Warriner, S. L. (1997). *J. Chem. Soc., Perkin Trans. 1*, 351.
57. Hense, A., Ley, S. V., Osborn, H. M. I., Owen, D. R., Poisson, J.-F., Warriner, S. L. and Wesson, K. E. (1997). *J. Chem. Soc., Perkin Trans. 1*, 2023.
58. Yin, H., Franck, R. W., Chen, S.-L., Quigley, G. J. and Todaro, L. (1992). *J. Org. Chem.*, **57**, 644.
59. Wander, J. D. and Horton, D. (1976). *Adv. Carbohydr. Chem. Biochem.*, **32**, 15.
60. Fischer, E. (1894). *Ber. Dtsch. Chem. Ges.*, **27**, 673.
61. Funabashi, M., Arai, S. and Shinohara, M. (1999). *J. Carbohydr. Chem.*, **18**, 333.
62. Rodriguez, E. B., Scally, G. D. and Stick, R. V. (1990). *Aust. J. Chem.*, **43**, 1391.
63. Miljković, M., Dropkin, D., Hagel, P. and Habash-Marino, M. (1984). *Carbohydr. Res.*, **128**, 11.
64. McAuliffe, J. C. and Hindsgaul, O. (1998). *Synlett*, 307.
65. Ferrier, R. J. and Furneaux, R. H. (1976). *Carbohydr. Res.*, **52**, 63.
66. Kartha, K. P. R. and Field, R. A. (1998). *J. Carbohydr. Chem.*, **17**, 693.
67. David, S. and Hanessian, S. (1985). *Tetrahedron*, **41**, 643.
68. David, S. (1997). Selective *O*-substitution and oxidation using atannylene acetals and stannyl ethers, in *Preparative Carbohydrate Chemistry*, Hanessian, S. ed., Marcel Dekker, New York, p. 69.
69. Grindley, T. B. (1998). *Adv. Carbohydr. Chem. Biochem.*, **53**, 17.
70. Munavu, R. M. and Szmant, H. H. (1976). *J. Org. Chem.*, **41**, 1832.
71. Nashed, M. A. and Anderson, L. (1976). *Tetrahedron Lett.*, 3503.
72. David, S., Thieffry, A. and Veyrières, A. (1981). *J. Chem. Soc., Perkin Trans. 1*, 1796.

73. Nikrad, P. V., Beierbeck, H. and Lemieux, R. U. (1992). *Can. J. Chem.*, **70**, 241.
74. Jenkins, D. J. and Potter, B. V. L. (1994). *Carbohydr. Res.*, **265**, 145.
75. Boons, G.-J., Castle, G. H., Clase, J. A., Grice, P., Ley, S. V. and Pinel, C. (1993). *Synlett*, 913.
76. Guilbert, B., Davis, N. J. and Flitsch, S. L. (1994). *Tetrahedron Lett.*, **35**, 6563.
77. Lubineau, A. and Lemoine, R. (1994). *Tetrahedron Lett.*, **35**, 8795.
78. Hodosi, G. and Kováč, P. (1998). *J. Carbohydr. Chem.*, **17**, 557.
79. Ogawa, T. and Matsui, M. (1981). *Tetrahedron*, **37**, 2363.
80. Alais, J. and Veyrières, A. (1981). *J. Chem. Soc., Perkin Trans. 1*, 377.
81. Dasgupta, F. and Garegg, P. J. (1994). *Synthesis*, 1121.

Amines

So far, the carbohydrates that we have encountered consist of just carbon, hydrogen and oxygen. However, other heteroatoms, nitrogen and phosphorus in particular, are commonly included in carbohydrate structures and an important class is that of the "amino sugars":

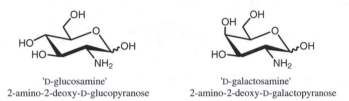

'D-glucosamine'
2-amino-2-deoxy-D-glucopyranose

'D-galactosamine'
2-amino-2-deoxy-D-galactopyranose

In the context of protecting groups, it seems appropriate to conclude the discussion with methods currently available for the protection of the amino group.

Traditionally, amino group protection in carbohydrates has relied on the chemistry developed earlier in the peptide field (benzyloxycarbonyl, *tert*-butoxycarbonyl). However, the removal of such carbamoyl groups requires the use of hydrogen or anhydrous acid, conditions that may adversely affect a carbohydrate, protected or otherwise. So arose the need to develop other protecting groups for the primary amine.

Nature has chosen her own such protecting group, the acetyl group: *N*-acetyl-D-glucosamine is a common component of many natural oligosaccharides and polysaccharides. When even this amide group is too reactive, chemists have been able to convert it into a rather benign "imide":[1,2]

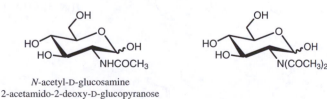

N-acetyl-D-glucosamine
2-acetamido-2-deoxy-D-glucopyranose

Two other imides are commonly used for the protection of the amino group, namely the phthalimide[3] and the tetrachlorophthalimide:[4-9]

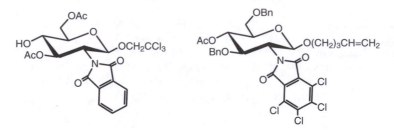

The latter was introduced because of its easier removal, compared to the phthalimide itself.

A not-so-obvious protection of the amino group is offered by the azide group; this group can be introduced at C2 of a carbohydrate by a number of means or, more spectacularly, by transformation of the amine itself.[10]

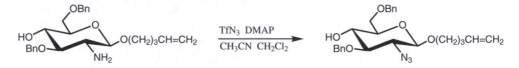

At a late stage in a synthesis, the azide group is easily reduced to reform the amine.

Two recent publications have suggested the 2,5-dimethylpyrrole and dimethylmaleoyl groups as useful amine protecting groups.[11,12] Finally, a summary of the reagents for the protection, and subsequent deprotection, of the amino group is given in Table 2.[13]

Table 2

	protection	deprotection	reference
$-NHAc \longrightarrow -NAc_2$	CH_3COCl, $EtNPr^i_2$	$NaOCH_3$, CH_3OH	1
	$CH_2C(CH_3)OAc$, TsOH		2
$-NH_2 \longrightarrow$		N_2H_4	3,14
		$H_2NCH_2CH_2NH_2$	15
$-NH_2 \longrightarrow$		$H_2NCH_2CH_2NH_2$	4,5
		$NaBH_4$, pH 5	6

Table 2 (*continued*)

	protection	deprotection	reference
$-N_3$ [6]		H_2, Pd-C Ph_3P, H_2O	3
$-NH_2 \longrightarrow -N_3$	TfN_3, DMAP		10
$-NH_2 \longrightarrow$	$CH_3CO(CH_2)_2COCH_3$, Et_3N	$NH_2OH \cdot HCl$	11
$-NH_2 \longrightarrow$		1. NaOH 2. H_3O^+	12

References

1. Castro-Palomino, J. C. and Schmidt, R. R. (1995). *Tetrahedron Lett.*, **36**, 6871.
2. Demchenko, A. V. and Boons, G.-J. (1998). *Tetrahedron Lett.*, **39**, 3065.
3. Banoub, J., Boullanger, P. and Lafont, D. (1992). *Chem. Rev.*, **92**, 1167.
4. Debenham, J. S., Madsen, R., Roberts, C. and Fraser-Reid, B. (1995). *J. Am. Chem. Soc.*, **117**, 3302.
5. Debenham, J. S., Debenham, S. D. and Fraser-Reid, B. (1996). *Bioorg. Med. Chem.*, **4**, 1909.
6. Castro-Palomino, J. C. and Schmidt, R. R. (1995). *Tetrahedron Lett.*, **36**, 5343.
7. Rodebaugh, R., Debenham, J. S. and Fraser-Reid, B. (1997). *J. Carbohydr. Chem.*, **16**, 1407.
8. Lergenmüller, M., Ito, Y. and Ogawa, T. (1998). *Tetrahedron*, **54**, 1381.
9. Olsson, L., Kelberlau, S., Jia, Z. J. and Fraser-Reid, B. (1998). *Carbohydr. Res.*, **314**, 273.
10. Olsson, L., Jia, Z. J. and Fraser-Reid, B. (1998). *J. Org. Chem.*, **63**, 3790.
11. Bowers, S. G., Coe, D. M. and Boons, G.-J. (1998). *J. Org. Chem.*, **63**, 4570.
12. Aly, M. R. E., Castro-Palomino, J. C., Ibrahim, E.-S. I., El-Ashry, E.-S. H. and Schmidt, R. R. (1998). *Eur. J. Org. Chem.*, 2305.
13. Jiao, H. and Hindsgaul, O. (1999). *Angew. Chem. Int. Ed.*, **38**, 346.
14. Bundle, D. R. and Josephson, S. (1979). *Can. J. Chem.*, **57**, 662.
15. Kanie, O., Crawley, S. C., Palcic, M. M. and Hindsgaul, O. (1993). *Carbohydr. Res.*, **243**, 139.

Chapter 7

The Reactions of Monosaccharides

So far, most of the chemical transformations performed on a monosaccharide have been concerned with the concept of protecting groups. In general, the order of reactivity of the various hydroxyl groups within D-glucopyranose was observed to be O1 (hemiacetal) >O6 (primary) >O2 (adjacent to Cl and therefore more acidic) >O3 >O4. Now, a whole host of chemical reactions will be discussed that cause functional group transformations within monosaccharides, either for the purpose of elaboration into oligosaccharides or for the synthesis of a derivative of the monosaccharide. Such derivatives include the deoxy and amino deoxy sugars.

By the nature of the subject, this discussion will be neither exhaustive nor all inclusive. It will, however, highlight the methods that work well in synthetic carbohydrate chemistry, particularly those that have evolved over the last few decades.

Oxidation and Reduction

We have already seen the importance of oxidation and reduction to the German chemists of the nineteenth century. Aldoses were oxidized to aldonic acids using bromine water and to aldaric acids using dilute nitric acid — these reagents are still used today. The reduction of an aldose to an alditol, however, is now more safely done with sodium borohydride rather than with the more toxic sodium amalgam.

Oxidation[1-3]

The major advances here have been concerned with the discovery or "invention", as some would call it,[a] of chemoselective reagents for the oxidation

---

[a] Sir Derek Barton co-authored a series of papers which highlighted the "invention" of new chemical reactions, particularly those concerned with free-radical pathways.[4]

of the various hydroxyl groups within a carbohydrate. The well established Jones' (CrO_3, H_2SO_4)[5] and Collins' (CrO_3, pyridine)[6] reagents soon gave way to milder and more easily handled oxidizing agents, namely pyridinium chlorochromate[7] and pyridinium dichromate.[8,9] These reagents oxidize both primary and secondary alcohols and some can produce either the aldehyde or the acid from a primary alcohol:[b]

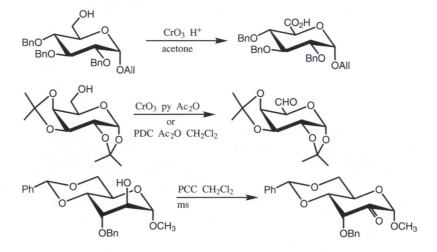

Another major class of oxidizing agents is based on dimethyl sulfoxide.[10] Since the initial report by Pfitzner and Moffatt (dimethyl sulfoxide, dicyclohexylcarbodiimide), there has been developed a string of reagents that activate the dimethyl sulfoxide for oxidation of, in particular, secondary alcohols:

$$(CH_3)_2SO \; + \; X^+ \; \longrightarrow \; (CH_3)_2S^+OX \; \xrightarrow[-HOX]{\overset{\backslash}{/}CHOH} \; H\text{-}\underset{|}{\overset{|}{C}}\text{-}O\text{-}\underset{CH_3}{\overset{CH_3}{S}}\text{+} \; \xrightarrow{-H^+}$$

$$\overset{\backslash}{\underset{/}{C}}=O \; + \; (CH_3)_2S$$

With primary alcohols, the formation of "(methylthio)methyl ethers" (actually thioacetals) can be troublesome:

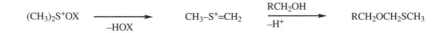

$$(CH_3)_2S^+OX \; \xrightarrow{-HOX} \; CH_3\text{-}S^+\text{=}CH_2 \; \xrightarrow[-H^+]{RCH_2OH} \; RCH_2OCH_2SCH_3$$

---

[b] From now on, structural formulae will utilize the abbreviations that are listed at the front of the book.

Some of the most popular activating agents for the dimethyl sulfoxide are acetic anhydride, trifluoroacetic anhydride, phosphorus pentaoxide, pyridine–sulfur trioxide and oxalyl chloride:[11,12]

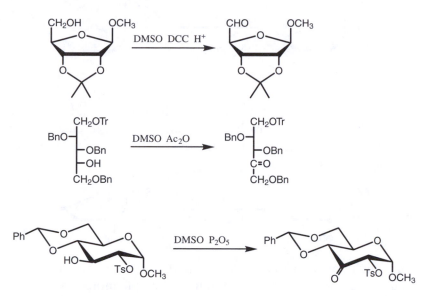

In fact, the activation of dimethyl sulfoxide by oxalyl chloride is such a general, mild and selective method of oxidation that it has taken on a mantle of its own — it is known as the "Swern" oxidation, after its founder:[13]

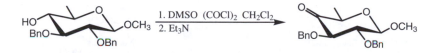

There has been a suggestion for the use of a recyclable sulfoxide, even polymer bound, to avoid the malodorous dimethyl sulfide that is liberated from oxidations conducted with dimethyl sulfoxide.[14]

A named reagent, the "Dess–Martin periodinane", has proven its value in general synthesis and must soon be used more regularly in the carbohydrate area:[15,16]

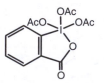

Ruthenium tetraoxide is a powerful oxidizing agent, the action of which can be moderated somewhat by employing small amounts of ruthenium(III) or ruthenium(IV) salts and stoichiometric amounts of another oxidizing agent:[17-19]

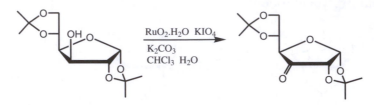

Ley, again, offered a novel variant of ruthenium-based oxidants by showing that tetrapropylammonium perruthenate could be used in catalytic amounts, together with another oxidizing agent, for the mild conversion of alcohols into aldehydes and ketones:[20]

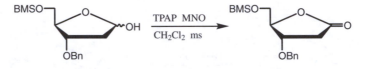

Two significant improvements to the method have been recently reported: a polymer-supported perruthenate and the use of simple molecular oxygen as the co-oxidant.[21]

2,2,6,6-Tetramethylpiperidine-1-oxyl, used in catalytic amounts with a co-oxidant, is capable of oxidizing primary alcohols to either the aldehyde or the carboxylic acid:[22-24]

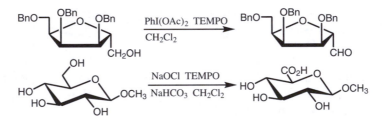

The latter example illustrates the utility of the method that seems to compare favourably with the more traditional oxygen-over-platinum dehydrogenation[25] for producing uronic acids. The mechanism of action of this interesting oxidant is thought to involve an oxoammonium ion:[26]

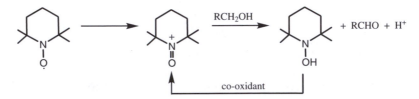

In connection with the work on the reaction of vicinal diols with bis(tributyltin) oxide and with dibutyltin oxide, it was found that the resulting tin-containing derivatives could be oxidized rapidly and selectively by molecular bromine or by *N*-bromosuccinimide:[27–30]

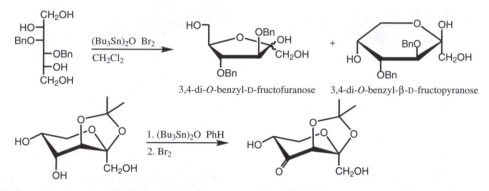

3,4-di-*O*-benzyl-D-fructofuranose   3,4-di-*O*-benzyl-β-D-fructopyranose

This oxidation is truly remarkable, not only in its activation of both the substrate (diol) and the reagent (bromine) but also in the regiochemistry displayed.

It is probably worth noting at this stage that many of the ketones produced by the oxidation of carbohydrate secondary alcohols are often found as stable, well defined hydrates — presumably, the numerous electronegative oxygen atoms present in the structure raise the reactivity of the carbonyl group. Hydrogen bonding in these hydrates may also be a stabilizing factor.[18]

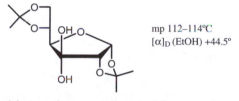

mp 112–114°C
$[\alpha]_D$ (EtOH) +44.5°

This section would not be complete without reference to the oxidative cleavage of carbohydrate vicinal diols by reagents such as periodic acid and lead(IV) acetate. Some fifty years ago, these sorts of oxidations not only provided valuable information to do with relative configuration and ring size[31] but also were of synthetic utility:[32]

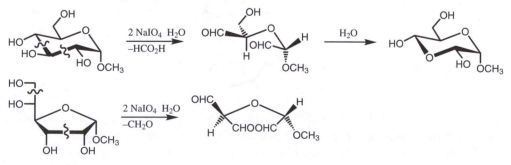

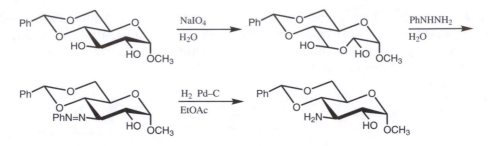

These days, of course, the problems of ring size and relative configuration are much more easily solved by nuclear magnetic resonance spectroscopy.[33–36]

Reduction

For many years, chemists were restricted to just "catalytic hydrogenation" and "dissolving metals" for the reduction of carbohydrates. Still, many interesting reactions were reported:[37–39]

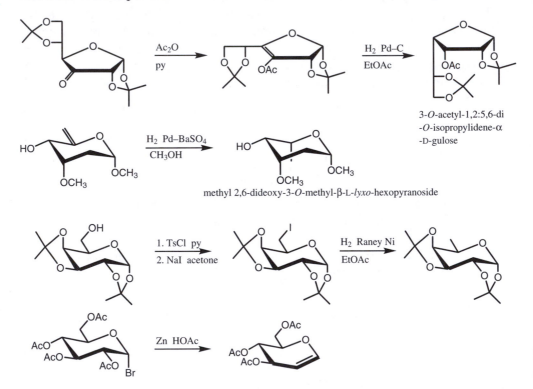

3-O-acetyl-1,2:5,6-di -O-isopropylidene-α -D-gulose

methyl 2,6-dideoxy-3-O-methyl-β-L-lyxo-hexopyranoside

The concept of "catalytic transfer hydrogenation" has obviated a lot of the need for using molecular hydrogen in transformations such as the above.[40]

With the advent of sodium borohydride and lithium aluminium hydride, many of the reductions performed with carbohydrates became easier and, with sodium borohydride, safer. This safety aspect was no more important than in the Kiliani–Fischer synthesis where the sodium amalgam was replaced by sodium borohydride for the reductions of the lactone to the aldose:

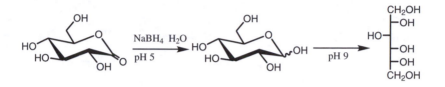

Other interesting transformations followed, some involving modified boro- and aluminium-hydrides:[18,41–43]

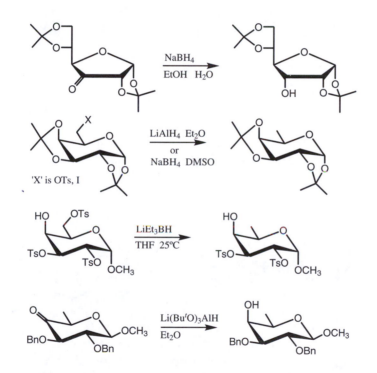

Tributyltin hydride has been used routinely for the removal of halogen from carbohydrate derivatives:[44,45]

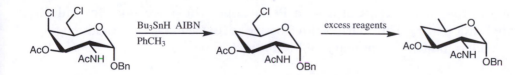

However, this reducing agent received a great boost in 1975 when Barton and McCombie reported the deoxygenation of secondary alcohols *via*, among other functionalities, the derived xanthates:[46,47]

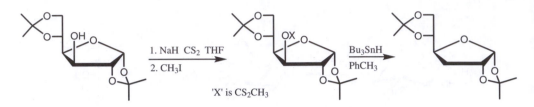

'X' is CS$_2$CH$_3$

This "invention" was much touted by Barton because of the free radical nature of the reaction:

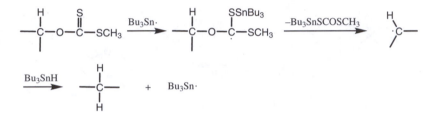

Over the years, the process has been refined and improved and is really the method of choice for the deoxygenation of a secondary alcohol.[48-50] Apart from the original xanthate, an equally attractive substrate is a thiocarbonate prepared under less basic conditions:[51]

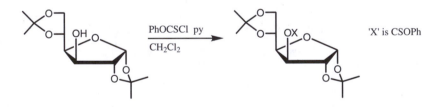

'X' is CSOPh

Although tin hydrides are still used routinely for the reduction, the waste products are highly toxic and pose a problem in disposal. A useful procedure has been reported for the removal of these toxic tin residues[52] and there exist methods that use a catalytic amount of the tin reagent and a stoichiometric amount of another reducing agent;[45] a polystyrene-supported organotin hydride has been developed.[53,54] However, it is felt that reagents based on a Si—H or, more probably in terms of cost, a P—H bond will prove safer for industrial application.[55-58] The Barton–McCombie deoxygenation method has been suitably modified for primary and tertiary alcohols.[50,57,59]

An extension of the above thiocarbonyl-based methodology can give rise to the selective deoxygenation of a diol:[60,61]

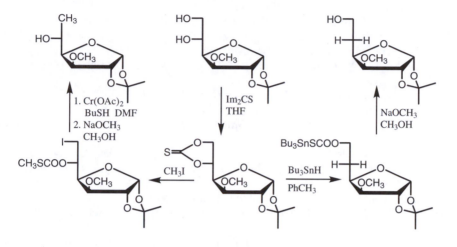

In summary, then, the best method for the deoxygenation of a primary alcohol probably still involves a conventional sequence:

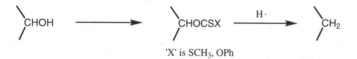

However, it is the Barton–McCombie free radical procedure that is best for secondary alcohols:

$$\text{CHOH} \longrightarrow \text{CHOCSX} \xrightarrow{\text{H·}} \text{CH}_2$$

'X' is SCH₃, OPh

References

1. Muzart, J. (1992). *Chem. Rev.*, **92**, 113.
2. Hudlický, M. (1990). *Oxidations in Organic Chemistry*, ACS Monograph 186, Washington DC, American Chemical Society.
3. Haines, A. H. (1988). *Methods for the Oxidation of Organic Compounds*, Academic Press, London.
4. Barton, D. H. R. and Liu, W. (1997). *Tetrahedron*, **53**, 12067.
5. van Boeckel, C. A. A., Delbressine, L. P. C. and Kaspersen, F. M. (1985). *Recl. Trav. Chim. Pays-Bas*, **104**, 259.
6. Garegg, P. J. and Samuelsson, B. (1978). *Carbohydr. Res.*, **67**, 267.
7. Piancatelli, G., Scettri, A. and D'Auria, M. (1982). *Synthesis*, 245.

8. Herscovici, J., Egron, M.-J. and Antonakis, K. (1982). *J. Chem. Soc., Perkin Trans. 1*, 1967.
9. Andersson, F. and Samuelsson, B. (1984). *Carbohydr. Res.*, **129**, C1.
10. Tidwell, T. T. (1990). *Synthesis*, 857.
11. Jones, G. H. and Moffatt, J. G. (1972). *Methods Carbohydr. Chem.*, **6**, 315.
12. Lindberg, B. (1972). *Methods Carbohydr. Chem.*, **6**, 323.
13. Mancuso, A. J. and Swern, D. (1981). *Synthesis*, 165.
14. Harris, J. M., Liu, Y., Chai, S., Andrews, M. D. and Vederas, J. C. (1998). *J. Org. Chem.*, **63**, 2407.
15. Meyer, S. D. and Schreiber, S. L. (1994). *J. Org. Chem.*, **59**, 7549.
16. Speicher, A., Bomm, V. and Eicher, T. (1996). *J. prakt. Chem.*, **338**, 588.
17. Haines, A. H. (1988). *Methods for the Oxidation of Organic Compounds*, Academic Press, London, p. 54.
18. Baker, D. C., Horton, D. and Tindall, C. G., Jr. (1972). *Carbohydr. Res.*, **24**, 192.
19. Morris, P. E., Jr., Kiely, D. E. and Vigee, G. S. (1990). *J. Carbohydr. Chem.*, **9**, 661.
20. Ley, S. V., Norman, J., Griffith, W. P. and Marsden, S. P. (1994). *Synthesis*, 639.
21. Hinzen, B., Lenz, R. and Ley, S. V. (1998). *Synthesis*, 977.
22. De Mico, A., Margarita, R., Parlanti, L., Vescovi, A. and Piancatelli, G. (1997). *J. Org. Chem.*, **62**, 6974.
23. Melvin, F., McNeill, A., Henderson, P. J. F. and Herbert, R. B. (1999). *Tetrahedron Lett.*, **40**, 1201.
24. Dijksman, A., Arends, I. W. C. E. and Sheldon, R. A. (2000). *Chem. Commun.*, 271.
25. Haines, A. H. (1988). *Methods for the Oxidation of Organic Compounds*, Academic Press, London, p. 10.
26. de Nooy, A. E. J., Besemer, A. C. and van Bekkum, H. (1996). *Synthesis*, 1153.
27. Zou, W. and Szarek, W. A. (1994). *Carbohydr. Res.*, **254**, 25.
28. den Drijver, L., Holzapfel, C. W., Koekemoer, J. M., Kruger, G. J. and van Dyk, M. S. (1986). *Carbohydr. Res.*, **155**, 141.
29. Kong, X. and Grindley, T. B. (1993). *J. Carbohydr. Chem.*, **12**, 557.
30. Hanessian, S. and Roy, R. (1985). *Can. J. Chem.*, **63**, 163.
31. Perlin, A. S. (1980). Glycol-cleavage oxidation, in *The Carbohydrates*, Pigman, W. and Horton, D. eds, vol. IB, Academic Press, London, p. 1167.
32. Patroni, J. J. and Stick, R. V. (1985). *Aust. J. Chem.*, **38**, 947.
33. Pigman, W. and Horton, D. eds (1980). *The Carbohydrates*, vol. IB, Academic Press, London, p. 1299.
34. Gorin, P. A. J. (1981). *Adv. Carbohydr. Chem. Biochem.*, **38**, 13.
35. Tvaroska, I. and Taravel, F. R. (1995). *Adv. Carbohydr. Chem. Biochem.*, **51**, 15.
36. Dais, P. and Perlin, A. S. (1987). *Adv. Carbohydr. Chem. Biochem.*, **45**, 125.
37. Lemieux, R. U. and Stick, R. V. (1975). *Aust. J. Chem.*, **28**, 1799.
38. Monneret, C., Conreur, C. and Khuong-Huu, Q. (1978). *Carbohydr. Res.*, **65**, 35.
39. Roth, W. and Pigman, W. (1963). *Methods Carbohydr. Chem.*, **2**, 405.
40. Johnstone, R. A. W., Wilby, A. H. and Entwistle, I. D. (1985). *Chem. Rev.*, **85**, 129.
41. Thiem, J. and Meyer, B. (1980). *Chem. Ber.*, **113**, 3067.
42. Binkley, R. W. (1985). *J. Org. Chem.*, **50**, 5646.
43. McAuliffe, J. C. and Stick, R. V. (1997). *Aust. J. Chem.*, **50**, 197.
44. Neumann, W. P. (1987). *Synthesis*, 665.
45. Baguley, P. A. and Walton, J. C. (1998). *Angew. Chem. Int. Ed.*, **37**, 3072.
46. Barton, D. H. R. and McCombie, S. W. (1975). *J. Chem. Soc., Perkin Trans. 1*, 1574.

47. Lee, A. W. M., Chan, W. H., Wong, H. C. and Wong, M. S. (1989). *Synth. Commun.*, **19**, 547.
48. Hartwig, W. (1983). *Tetrahedron*, **39**, 2609.
49. Barton, D. H. R. (1992). *Tetrahedron*, **48**, 2529.
50. Barton, D. H. R., Ferreira, J. A. and Jaszberenyi, J. C. (1997). Free radical deoxygenation of thiocarbonyl derivatives of alcohols, in *Preparative Carbohydrate Chemistry*, Hanessian, S. ed., New York, Marcel Dekker, p. 151.
51. Robins, M. J., Wilson, J. S. and Hansske, F. (1983). *J. Am. Chem. Soc.*, **105**, 4059.
52. Berge, J. M. and Roberts, S. M. (1979). *Synthesis*, 471.
53. Neumann, W. P. and Peterseim, M. (1992). *Synlett*, 801.
54. Boussaguet, P., Delmond, B., Dumartin, G. and Pereyre, M. (2000). *Tetrahedron Lett.*, **41**, 3377.
55. Lesage, M., Chatgilialoglu, C. and Griller, D. (1989). *Tetrahedron Lett.*, **30**, 2733.
56. Schummer, D. and Höfle, G. (1990). *Synlett*, 705.
57. Jang, D. O., Cho, D. H. and Barton, D. H. R. (1998). *Synlett*, 39.
58. Barton, D. H. R. and Jacob, M. (1998). *Tetrahedron Lett.*, **39**, 1331.
59. Barton, D. H. R., Blundell, P., Dorchak, J., Jang, D. O. and Jaszberenyi, J. Cs. (1991). *Tetrahedron*, **47**, 8969.
60. Barton, D. H. R. and Stick, R. V. (1975). *J. Chem. Soc., Perkin Trans. 1*, 1773.
61. Barton, D. H. R. and Subramanian, R. (1977). *J. Chem. Soc., Perkin Trans. 1*, 1718.

Halogenation

We have just seen that deoxyhalogeno sugars were of some importance in the synthesis of deoxy sugars. Although all of the halogens could be introduced into a monosaccharide using the older methods and reagents, there were some limitations as to what could be achieved:[1]

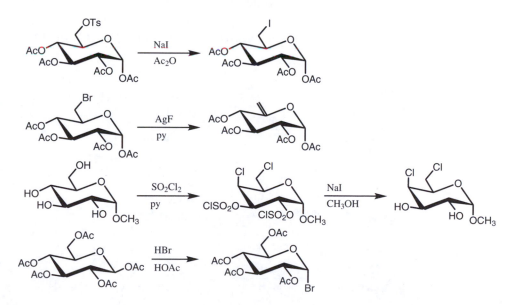

It soon became obvious that additional processes were needed for the introduction of halogen at the lone primary carbon, compared to any of the three secondary carbon atoms. In addition, milder conditions could be an advantage for the introduction of halogen at the anomeric carbon; the product, a glycosyl halide, was exceptionally reactive, being an "α-halo ether".

There was naturally a need for improved methods and, perhaps, the aforementioned Hanessian–Hullar reaction was one of these:

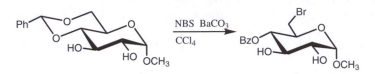

Many of the reagents for the introduction of halogen now rely on the molecule, triphenylphosphine. The discussion that follows will, by necessity, separate the introduction of halogen at the anomeric and non-anomeric positions.

Non-anomeric Halogenation

Although the traditional nucleophilic displacements, for example, of a tosylate by iodide, are still used, the methods currently in vogue for the introduction of chlorine, bromine and iodine all utilize triphenylphosphine. Probably the most popular choices are triphenylphosphine with carbon tetrachloride, carbon tetrabromide and iodine/imidazole. In fact, this last combination has become the reagent of choice for the regioselective iodination (at C6) of many hexopyranosides:[1–5]

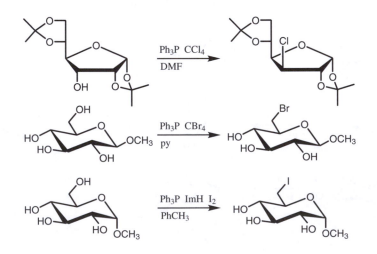

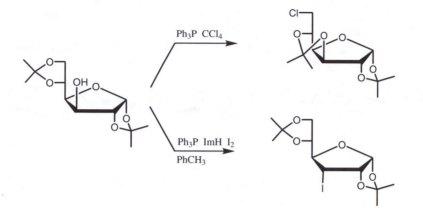

Where appropriate, it can be seen that the above halogenations proceed with inversion of configuration at carbon. This observation is in line with a general principle of nucleophilic displacement reactions in carbohydrates — the preference for S_N2 processes over S_N1. Presumably, carbenium ion intermediates are disfavoured by the presence of so many (electronegative) oxygen atoms. A general rationalisation of triphenylphosphine-mediated halogenation follows, with the driving force being the formation of the very stable phosphine oxide:

$$Ph_3P + E^+ \longrightarrow Ph_3\overset{+}{P}E \underset{-HE}{\overset{\backslash CHOH}{\longrightarrow}} \overset{\backslash}{\underset{/}{C}}HO\overset{+}{P}Ph_3 \overset{X^-}{\longrightarrow}$$

$$X\overset{/}{\underset{\backslash}{C}}H + Ph_3PO$$

Although fluorine can be introduced into a carbohydrate by conventional nucleophilic substitution, often employing a sulfonate and tetrabutylammonium fluoride, the method of choice undoubtedly employs (diethylamino)sulfur trifluoride:[6-8]

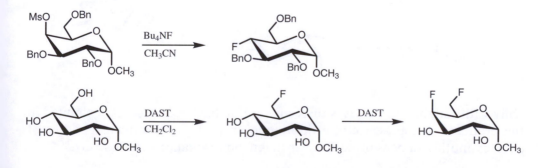

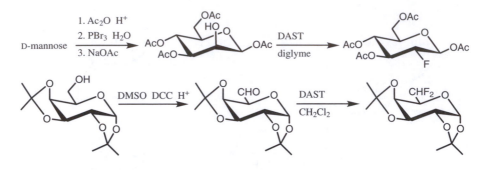

The reagent is relatively expensive but reliable and the overall fluorine substitution again proceeds with inversion of configuration:

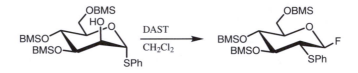

Because of the high reactivity of (diethylamino)sulfur trifluoride, rearrangements are often observed in susceptible substrates:[9,10]

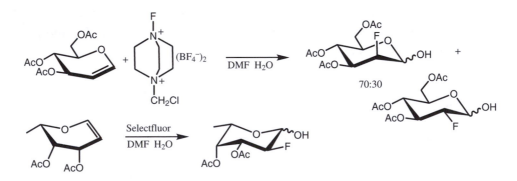

A novel preparation of 2-deoxy-2-fluoro sugars involves the addition of "Selectfluor" {1-chloromethyl-4-fluoro-1,4-diazoniabicyclo[2.2.2]octane bis(tetrafluoroborate)} to a 1,2-alkene:[8,11–16]

Only with suitable substrates is there any useful diastereoselectivity exhibited in the products. The presence of a fluorine atom in deoxyfluoro sugars adds, of course, another dimension to nuclear magnetic resonance spectroscopy.[17]

Anomeric Halogenation

Glycosyl halides, as mentioned, can be viewed as α-halo ethers and, consequently, their method of preparation and subsequent display of chemical reactivity are different from the normal deoxyhalogeno sugars. The first examples of this valuable class were the glycosyl chlorides and bromides:[18−20]

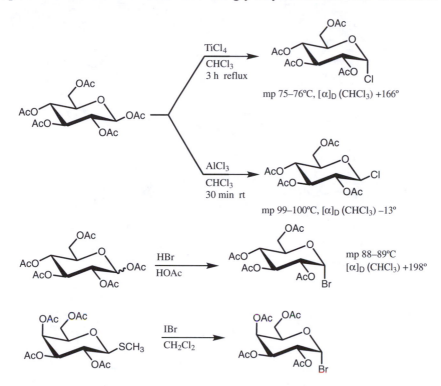

Generally, glycosyl chlorides can be obtained in both the α- and the β-D-pyranose forms. The α-anomer is naturally the more stable but the strength of the carbon-chlorine bond allows the formation of the β-anomer under conditions of kinetic control:

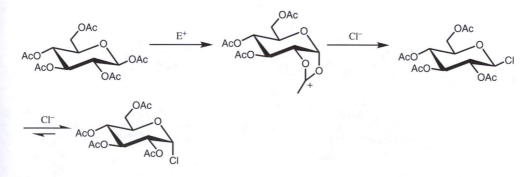

Glycosyl bromides, because of the weaker carbon–bromine bond, are generally isolated only as the α-D-anomer. This also holds for glycosyl iodides; in fact, until recently, glycosyl iodides were formed only rarely and were deemed too unstable to be of any use in subsequent synthetic transformations:[19,21–26]

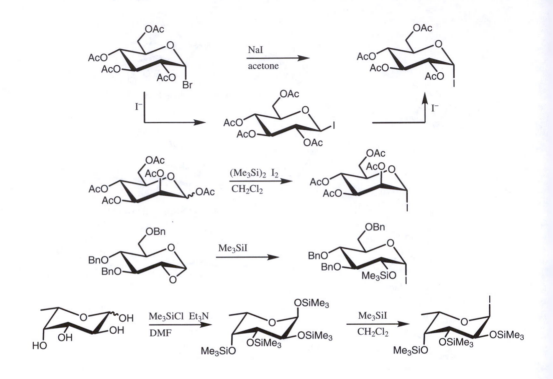

Glycosyl fluorides have been known for many years but it took the inventiveness of chemists such as Mukaiyama and Nicolaou to realize the potential of these fluorides and so stimulate new methods for their preparation:[6,27,28]

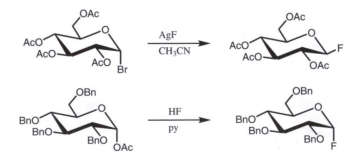

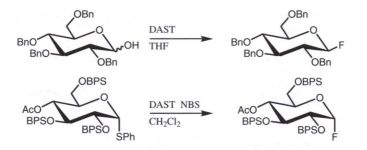

The α-D-anomer is the more stable product but, again, owing to the strength of the carbon–fluorine bond, it is possible to prepare the β-D-anomer in high yield. Another noteworthy feature of glycosyl fluorides is that, uniquely, they may be deprotected to give the polyol:[6]

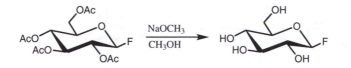

Some interesting and useful molecules result from the combination of various methods:

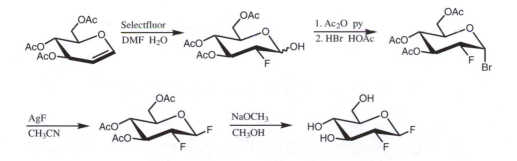

Recently, it has been shown that all of the glycosyl halides can be prepared from the free sugar, using haloenamine reagents:[29]

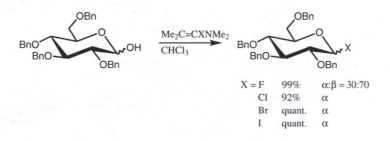

X = F	99%	α:β = 30:70
Cl	92%	α
Br	quant.	α
I	quant.	α

A very recent report describes a convenient synthesis of glycosyl halides, again from the free sugar:[30]

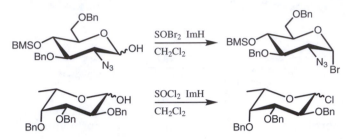

Finally, mention will be made of a reaction that does not occur at the anomeric centre but actually generates *another*, additional anomeric centre — Ferrier's "photobromination":[31]

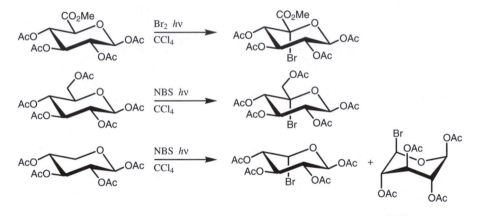

The brominated compounds can be put to good synthetic use:[31,32]

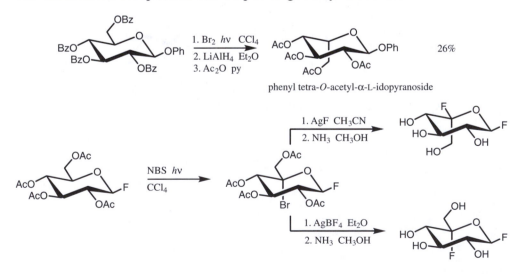

phenyl tetra-*O*-acetyl-α-L-idopyranoside

These glycosyl fluorides, when 2- and 5-fluoro substituted, are of invaluable use in the labelling of amino acid residues located in the active site of various glycoside hydrolases.[33,34]

References

1. Szarek, W. A. (1973). *Adv. Carbohydr. Chem. Biochem.*, **28**, 225.
2. Castro, B. R. (1983). *Org. React.*, **29**, 1.
3. Haylock, C. R., Melton, L. D., Slessor, K. N. and Tracey, A. S. (1971). *Carbohydr. Res.*, **16**, 375.
4. Whistler, R. L. and Anisuzzaman, A. K. M. (1980). *Methods Carbohydr. Chem.*, **8**, 227.
5. Garegg, P. J. (1984). *Pure Appl. Chem.*, **56**, 845.
6. Tsuchiya, T. (1990). *Adv. Carbohydr. Chem. Biochem.*, **48**, 91.
7. Taylor, N. F. ed. (1988). *Fluorinated Carbohydrates. Chemical and Biochemical Aspects*, ACS Symposium Series 374, American Chemical Society: Washington DC.
8. Dax, K., Albert, M., Ortner, J. and Paul, B. J. (1999). *Current Org. Chem.*, **3**, 287.
9. Nicolaou, K. C., Ladduwahetty, T., Randall, J. L. and Chucholowski, A. (1986). *J. Am. Chem. Soc.*, **108**, 2466.
10. Barrena, M. I., Matheu, M. I. and Castillón, S. (1998). *J. Org. Chem.*, **63**, 2184.
11. Banks, R. E. (1998). *J. Fluorine Chem.*, **87**, 1.
12. Vincent, S. P., Burkart, M. D., Tsai, C.-Y., Zhang, Z. and Wong, C.-H. (1999). *J. Org. Chem.*, **64**, 5264.
13. Albert, M., Dax, K. and Ortner, J. (1998). *Tetrahedron*, **54**, 4839.
14. Ortner, J., Albert, M., Weber, H. and Dax, K. (1999). *J. Carbohydr. Chem.*, **18**, 297.
15. Taylor, S. D., Kotoris, C. C. and Hum, G. (1999). *Tetrahedron*, **55**, 12431.
16. Lamberth, C. (1998). *Recent Res. Dev. Synth. Org. Chem.*, 1.
17. Csuk, R. and Glänzer, B. I. (1988). *Adv. Carbohydr. Chem. Biochem.*, **46**, 73.
18. Lemieux, R. U. (1963). *Methods Carbohydr. Chem.*, **2**, 223, 224.
19. Haynes, L. J. and Newth, F. H. (1955). *Adv. Carbohydr. Chem.*, **10**, 207.
20. Kartha, K. P. R. and Field, R. A. (1997). *Tetrahedron Lett.*, **38**, 8233.
21. Kartha, K. P. R. and Field, R. A. (1998). *Carbohydr. Lett.*, **3**, 179.
22. Gervay, J. (1998). Glycosyl iodides in organic synthesis, in *Organic Synthesis: Theory and Applications*, Hudlicky, T. ed., vol. 4, JAI Press, Stamford CT, p. 121.
23. Hadd, M. J. and Gervay, J. (1999). *Carbohyhdr. Res.*, **320**, 61.
24. Uchiyama, T. and Hindsgaul, O. (1996). *Synlett*, 499.
25. Uchiyama, T. and Hindsgaul, O. (1998). *J. Carbohydr. Chem.*, **17**, 1181.
26. Caputo, R., Kunz, H., Mastroianni, D., Palumbo, G., Pedatella, S. and Solla, F. (1999). *Eur. J. Org. Chem.*, 3147.
27. Shimizu, M., Togo, H. and Yokoyama, M. (1998). *Synthesis*, 799.
28. Miethchen, R. and Defaye, J. eds. (2000). *Carbohydr. Res.*, **327**, 1–218.
29. Ernst, B. and Winkler, T. (1989). *Tetrahedron Lett.*, **30**, 3081.
30. Baeschlin, D. K., Chaperon, A. R., Charbonneau, V., Green, L. G., Ley, S. V., Lücking, U. and Walther, E. (1998). *Angew. Chem. Int. Ed.*, **37**, 3423.
31. Somsák, L. and Ferrier, R. J. (1991). *Adv. Carbohydr. Chem. Biochem.*, **49**, 37.
32. Wong, A. W., He, S. and Withers, S. G. *Can. J. Chem.*, in press.

33. Davies, G. J., Mackenzie, L., Varrot, A., Dauter, M., Brzozowski, A. M., Schülein, M. and Withers, S. G. (1998). *Biochemistry*, **37**, 11707.
34. Howard, S., He, S. and Withers, S. G. (1998). *J. Biol. Chem.*, **273**, 2067.

Alkenes and Carbocycles

It is appropriate to discuss alkenes and carbocycles together as the former are sometimes precursors to the latter. In general, alkenes derived from carbohydrates are rarely found naturally and usually result from some chemical transformation in the laboratory, for example, dehydration, dehydrohalogenation or "de-dihydroxylation". Again, deoxy anomeric alkenes, termed "glycals", are unique and their chemistry will be discussed separately.

Carbocycles form the core of several classes of biologically important molecules, for example, the cyclitols. It is fitting that chemists have mimicked Nature somewhat in being able to convert a monosaccharide alkene into a carbocycle.

Alkenes[1]

Non-anomeric alkenes: In the pyranose series, 2,3-, 3,4-, 4,5- and 5,6-alkenes are all known, with the last type probably being the most common and useful. The generation of a 2,3- or 3,4-alkene can conveniently occur from the appropriate diol, using older (Tipson–Cohen) methodology or an extension of methods that we have already seen as developed by Barton and by Garegg:[2–5]

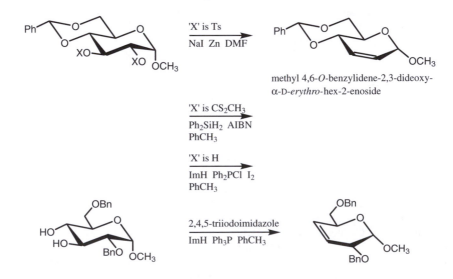

methyl 4,6-*O*-benzylidene-2,3-dideoxy-
α-D-*erythro*-hex-2-enoside

A rather unusual method for the synthesis of a 2,3-alkene involves the reduction of a cyclic sulfate:[6]

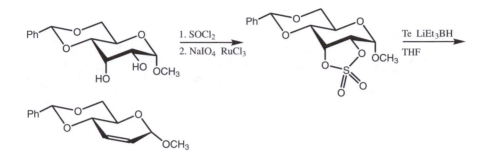

For the synthesis of a 4,5- or 5,6-alkene, a dehydration process may appear obvious but it is a dehydrohalogenation that is more efficient:[7]

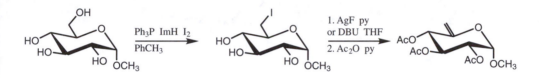

An unusual synthesis of acyclic 5,6-alkenes involves the reduction of a 6-bromo-6-deoxy-D-glucopyranoside:[8]

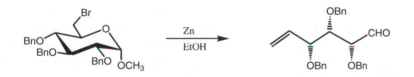

Anomeric alkenes: Although 1,5-anhydro-hex-1-enitols are known, it is the 2-deoxy derivatives which are much more common and synthetically useful; the name of "glycal" for the latter group was given when the properties of an aldehyde (from hydrolysis of the enol structure) were originally noticed:

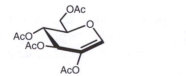

3,4,6-tri-O-acetyl-1,5-anhydro-2-deoxy-D-*arabino*-hex-1-enitol
(D-glucal triacatate)

There are many methods for the formation of glycals, ranging from the original treatment of "acetobromglucose" with zinc in acetic acid (by Fischer in 1913) to a comparable reduction with samarium(II) iodide:[1,9–13]

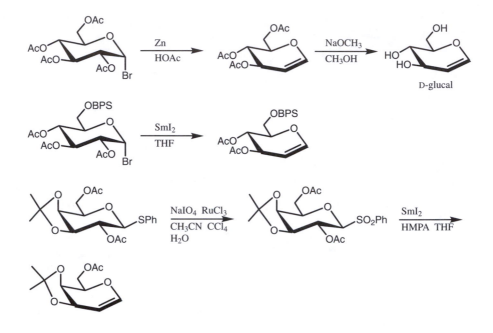

The use of zinc/silver/graphite offers a significant improvement[14] and a "one pot" procedure seems to be of advantage when conversion of a free sugar directly into the glycal is required.[15] An aprotic procedure (zinc and a tertiary amine in benzene) has also been developed[16] and potassium–graphite laminate is a vigorous enough reagent to cause the formation of glycals even from *O*-alkyl or *O*-alkylidene precursors:[17]

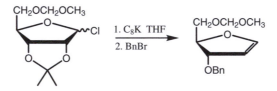

Furanoid glycals are generally difficult to prepare and are much less stable than their pyranoid counterparts. New methods employing other metals, for example, titanium and cobalt (of vitamin B_{12}) continue to be announced.[18–20]

The second type of anomeric alkene is one that is acyclic and generated by the Wittig reaction, or one of its well-known variants:[21]

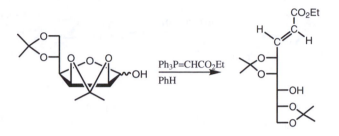

It is this very sort of alkene that has found so much use in the synthesis of carbocycles.

Carbocycles[22,23]

For the biosynthesis of the inositols (polyhydroxylated cyclohexanes), Nature uses a carbohydrate precursor, D-*xylo*-hexos-5-ulose 6-phosphate, and makes use of the aldol reaction:[24]

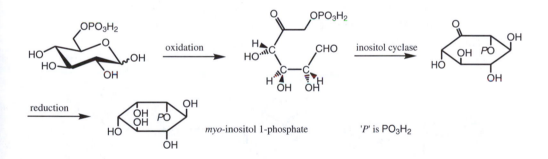

This sort of reaction has been mimicked[25] and emulated by chemists for the synthesis of carbocycles from carbohydrates and, without doubt, the procedure developed by Ferrier is the most valuable:[7,23,26]

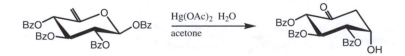

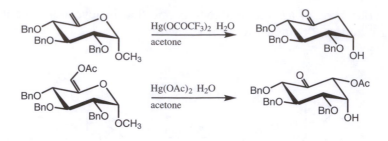

The generally accepted mechanism for this reaction involves an oxymercuration, followed by an intramolecular aldol reaction:[23]

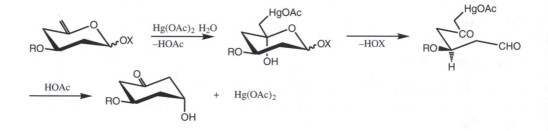

The stereochemistry that is observed in the final product has been commented upon.[23] Certainly, one of the strengths of this transformation is the good use to which the initial products can be put:[7,23,26–28]

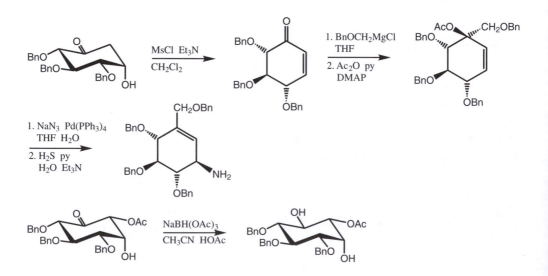

Two variations of this carbocycle synthesis, both of which retain the anomeric linkage, have recently been reported:[29]

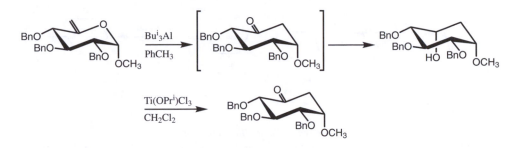

Another approach to the synthesis of carbocycles from carbohydrates involves one of the anomeric alkenes discussed previously, often utilizing free radical chemistry:[22,23,30]

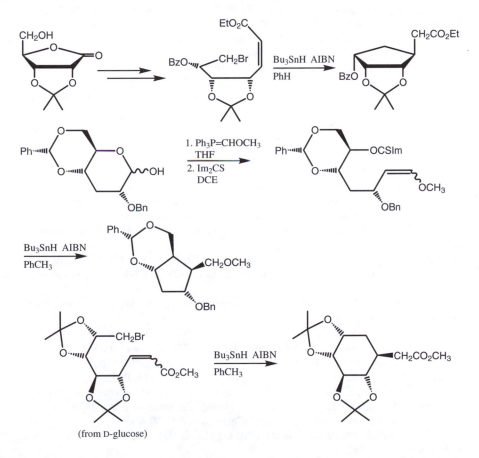

These sorts of reactions all involve free radical intermediates, with hex-5-enyl radicals cyclizing more rapidly than their hept-6-enyl counterparts and in accord with the rules as espoused by Baldwin and Beckwith:[22,31–34]

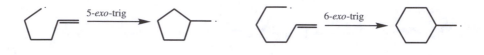

References

1. Ferrier, R. J. (1969). *Adv. Carbohydr. Chem. Biochem.*, **24**, 199.
2. Baptistella, L. H. B., Neto, A. Z., Onaga, H. and Godoi, E. A. M. (1993). *Tetrahedron Lett.*, **34**, 8407.
3. Barton, D. H. R., Jang, D. O. and Jaszberenyi, J. Cs. (1991). *Tetrahedron Lett.*, **32**, 2569.
4. Garegg, P. J. and Samuelsson, B. (1979). *Synthesis*, 813.
5. Liu, Z., Classon, B. and Samuelsson, B. (1990). *J. Org. Chem.*, **55**, 4273.
6. Chao, B., McNulty, K. C. and Dittmer, D. C. (1995). *Tetrahedron Lett.*, **36**, 7209.
7. McAuliffe, J. C. and Stick, R. V. (1997). *Aust. J. Chem.*, **50**, 193.
8. Bernet, B. and Vasella, A. (1979). *Helv. Chim. Acta*, **62**, 2400.
9. Helferich, B. (1952). *Adv. Carbohydr. Chem.*, **7**, 209.
10. Ferrier, R. J. (1965). *Adv. Carbohydr. Chem.*, **20**, 67.
11. Roth, W. and Pigman, W. (1963). *Methods Carbohydr. Chem.*, **2**, 405.
12. Yamaura, M., Sato, S., Hashimoto, N., Obise, H., Ikuta, N., Umemura, K. and Yoshimura, J., unpublished results.
13. de Pouilly, P., Chénedé, A., Mallet, J.-M. and Sinaÿ, P. (1992). *Tetrahedron Lett.*, **33**, 8065.
14. Csuk, R., Fürstner, A., Glänzer, B. I. and Weidmann, H. (1986). *J. Chem. Soc., Chem. Commun.*, 1149.
15. Shull, B. K., Wu, Z. and Koreeda, M. (1996). *J. Carbohydr. Chem.*, **15**, 955.
16. Somsák, L. and Németh, I. (1993). *J. Carbohydr. Chem.*, **12**, 679.
17. Fürstner, A. and Weidmann, H. (1988). *J. Carbohydr. Chem.*, **7**, 773.
18. Forbes, C. L. and Franck, R. W. (1999). *J. Org. Chem.*, **64**, 1424.
19. Hansen, T., Krintel, S. L., Daasbjerg, K. and Skrydstrup, T. (1999). *Tetrahedron Lett.*, **40**, 6087.
20. Spencer, R. P., Cavallaro, C. L. and Schwartz, J. (1999). *J. Org. Chem.*, **64**, 3987.
21. Collins, P. M., Overend, W. G. and Shing, T. S. (1982). *J. Chem. Soc., Chem. Commun.*, 297.
22. Hanessian, S. ed. (1997). *Preparative Carbohydrate Chemistry*, Marcel Dekker, New York, pp. 543–591.
23. Ferrier, R. J. and Middleton, S. (1993). *Chem. Rev.*, **93**, 2779.
24. Barnett, J. E. G., Rasheed, A. and Corina, D. L. (1973). *Biochem. J.*, **131**, 21.
25. Kiely, D. E. and Sherman, W. R. (1975). *J. Am. Chem. Soc.*, **97**, 6810.
26. Bender, S. L. and Budhu, R. J. (1991). *J. Am. Chem. Soc.*, **113**, 9883.
27. Sato, K.-i., Sakuma, S., Muramatsu, S. and Bokura, M. (1991). *Chem. Lett.*, 1473.
28. Takahashi, H., Kittaka, H. and Ikegami, S. (1998). *Tetrahedron Lett.*, **39**, 9703.
29. Dalko, P. I. and Sinaÿ, P. (1999). *Angew. Chem. Int. Ed.*, **38**, 773.
30. Marco-Contelles, J., Pozuelo, C., Jimeno, M. L., Martínez, L. and Martínez-Grau, A. (1992). *J. Org. Chem.*, **57**, 2625.

31. Baldwin, J. E. (1976). *J. Chem. Soc., Chem. Commun.*, 734.
32. Beckwith, A. L. J., Easton, C. J. and Serelis, A. K. (1980). *J. Chem. Soc., Chem. Commun.*, 482.
33. Giese, B. (1986). *Radicals in Organic Synthesis: Formation of Carbon–Carbon Bonds*, Pergamon Press, Oxford.
34. Curran, D. P., Porter, N. A. and Giese, B. (1996). *Stereochemistry of Radical Reactions*, VCH, Weinheim.

Anhydro Sugars

As the name implies, anhydro sugars formally result from the loss of water between two hydroxyl groups within a monosaccharide unit, with the concomitant formation of a new carbon–oxygen bond. Because of the number of different types of hydroxyl group present in a monosaccharide, a whole range of anhydro sugars is possible; a classification based on whether the "anhydro linkage" involves the anomeric centre, or not, is convenient.

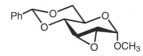

1,6-anhydro-β-D-glucopyranose   methyl 2,3-anhydro-4,6-*O*-benzylidene-α-D-alloside

Non-anomeric Anhydro Sugars[1]

The most important of the non-anomeric anhydro sugars are the epoxides. Such epoxides can be formed, somewhat circuitously, by the oxidation of an unsaturated sugar with some sort of peracid but, more commonly, they result from an internal nucleophilic displacement:[2-5]

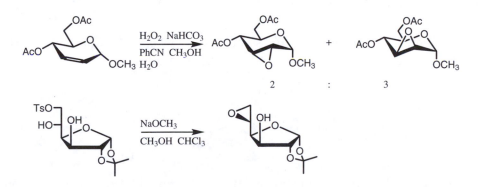

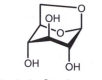

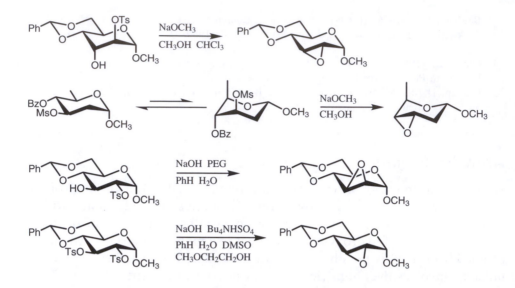

From the above, it can be seen that the direct oxidation of an alkene is (generally) not very stereoselective. In addition, the formation of an epoxide by the internal displacement of a sulfonic acid ester is a favoured process when either few conformational constraints exist or an "anti-periplanar" arrangement of nucleophile (O⁻) and leaving group (OTs, OMs) can be attained. For substrates where such an arrangement cannot be reached by the normal conformational change ($^4C_1$ to $^1C_4$), it has been suggested that other conformations may be involved.[1,6]

Another approach to sugar epoxides involves an intramolecular Mitsunobu reaction, directly on the diol:[6]

Care must be employed in some cases where an internal "Payne rearrangement" is possible in an initially formed product:[7,8]

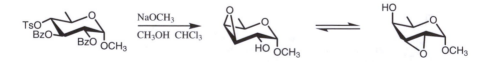

Spiro-epoxides, apart from arising from the epoxidation of an exocyclic alkene, can be formed from ketones and dimethylsulfoxonium methylide or dimethylsulfonium methylide:[9-11]

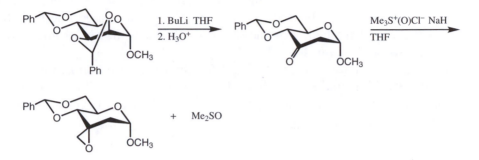

Of the other non-anomeric anhydro sugars, the formation of the 3,6-derivative is a common occurrence in the base treatment of glycosides with a good leaving group at C6:[12]

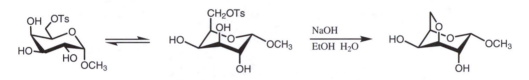

Anomeric Anhydro Sugars

Anomeric anhydro sugars, where the new carbon−oxygen bond is to the anomeric centre, may be considered as "internal" acetals. Again, many possibilities exist but, here, only the important 1,2- and 1,6-anhydro hexopyranoses will be discussed:

1,2-anhydro-3,4,6-tri-O-benzyl-α-D-glucose   1,6-anhydro-β-D-glucopyranose

1,6-Anhydro hexopyranoses:[12,13] Some aldohexoses, in aqueous acid at equilibrium, form a large amount of the 1,6-anhydro pyranose derivative:[15]

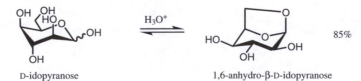

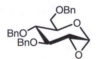

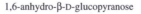

D-idopyranose   1,6-anhydro-β-D-idopyranose

For D-idose, as illustrated, the result is not surprising; the product is stabilized by a favourable anomeric effect and no longer suffers from having axial hydroxyl groups on the pyranose ring. In fact, for D-altrose, D-gulose and D-idose, this simple acid-catalysed process represents a viable method for preparing the anhydro sugar. For the other aldohexoses, however, alternative methods had to be developed.

One of the oldest ways of producing 1,6-anhydro hexopyranoses is by the pyrolysis of readily available carbohydrates, such as starch, cellulose, mannan and α-lactose. In this way are prepared 1,6-anhydro-β-D-glucopyranose, -mannopyranose and -galactopyranose:[16–18]

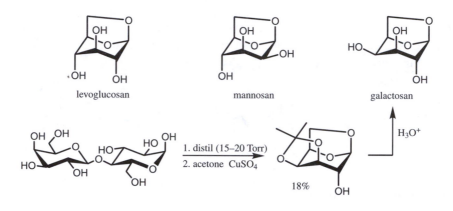

The trivial names arise from "glycosan" (*glycose an*hydride) and "levoglucosan" from the fact that an early report on glucosan gave a positive value for the optical rotation of what was obviously an impure sample.

Other methods are commonly used for the preparation of the glycosans, generally involving some sort of activation of either the C1–O1 or C6–O6 bond:[19–24]

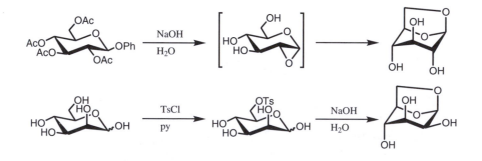

Glycals can be convenient sources of functionalized anhydro sugars:[25-29]

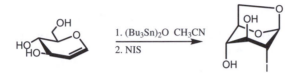

One of the reasons that so much effort has been spent on the synthesis of the glycosans is that they offer so much to the synthetic chemist in subsequent transformations. Included in the gift are dual protection of both C1 and C6, an unusual conformation ($^1C_4$), an array of hydroxyl groups of unusual and differing reactivity and an anhydro bridge which, when finally broken, allows a return to the normal $^4C_1$ conformation of the pyranose ring. Making use of most of these features are the "Cerny" 2,3- and 3,4-epoxides:[13,28-32]

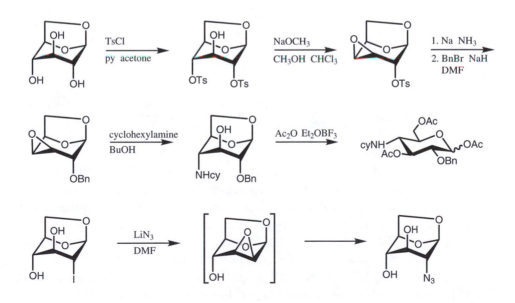

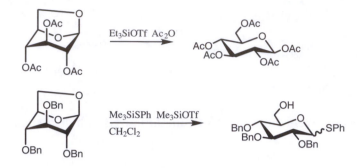

1,2-Anhydro hexopyranoses: The classic example of a 1,2-anhydro hexopyranose was by Brigl, many years ago:[33]

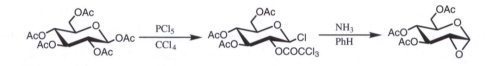

"Brigl's anhydride" took on a special role in the history of carbohydrate chemistry when it was used in the first chemical synthesis of sucrose by Lemieux.[34] Nowadays, this method of preparation of anomeric epoxides (the displacement of some good leaving group at C1 by an oxygen nucleophile at C2) has given way to more general procedures.

The most convenient and operationally simple method for the synthesis of 1,2-anhydro hexopyranoses involves the oxidation of glycals with dimethyl-dioxirane:[35,36]

As we have seen, all sorts of glycals are readily prepared and the reagent, dimethyldioxirane, is easily generated in acetone solution, free of moisture.[37] The work-up procedure involves nothing more than a simple evaporation of the solvent and, generally, one diastereoisomer is either formed exclusively or predominates. The use of more traditional oxidizing agents, for example, 3-chloroperbenzoic acid, produces acidic by-products that destroy the desired epoxide. More will be said about these anomeric epoxides later but, suffice it to say, they are generous precursors of glycosyl amines, acetals, thioacetals and glycosyl fluorides, as one would expect of an epoxide that is also part of an acetal:[38]

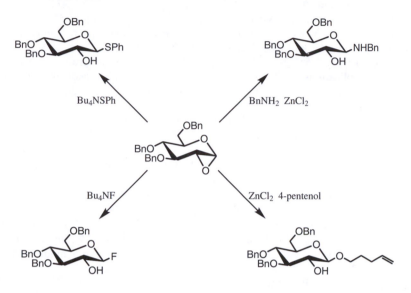

References

1. Williams, N. R. (1970). *Adv. Carbohydr. Chem. Biochem.*, **25**, 109.
2. Stewart, A. O. and Williams R. M. (1984). *Carbohydr. Res.*, **135**, 167.
3. Wiggins, L. F. (1963). *Methods Carbohydr. Chem.*, **2**, 188.
4. Szeja, W. (1986). *Carbohydr. Res.*, **158**, 245.
5. Szeja, W. (1988). *Carbohydr. Res.*, **183**, 135.
6. Guthrie, R. D. and Jenkins, I. D. (1981). *Aust. J. Chem.*, **34**, 1997.
7. Behrens, C. H., Ko, S. Y., Sharpless, K. B. and Walker, F. J. (1985). *J. Org. Chem.*, **50**, 5687.
8. Buchanan, J. G., Fletcher, R., Parry, K. and Thomas, W. A. (1969). *J. Chem. Soc. B*, 377.
9. Horton, D. and Weckerle, W. (1975). *Carbohydr. Res.*, **44**, 227.
10. King, R. D., Overend, W. G., Wells, J. and Williams, N. R. (1967). *J. Chem. Soc., Chem. Commun.*, 726.
11. Hagen, S., Anthonsen, T. and Kilaas, L. (1979). *Tetrahedron*, **35**, 2583.
12. Lewis, B. A., Smith, F. and Stephen, A. M. (1963). *Methods Carbohydr. Chem.*, **2**, 172.
13. Černý, M. and Staněk, J., Jr. (1977). *Adv. Carbohydr. Chem. Biochem.*, **34**, 23.
14. Bols, M. (1996). 1,6-Anhydro sugars, in *Carbohydrate Building Blocks*, John Wiley and Sons, New York, pp. 43–48.
15. Angyal, S. J. and Dawes, K. (1968). *Aust. J. Chem.*, **21**, 2747.
16. Ward, R. B. (1963). *Methods Carbohydr. Chem.*, **2**, 394.
17. Shafizadeh, F., Furneaux, R. H., Stevenson, T. T. and Cochran, T. G. (1978). *Carbohydr. Res.*, **67**, 433.
18. Hann, R. M. and Hudson, C. S. (1942). *J. Am. Chem. Soc.*, **64**, 2435.
19. Coleman, G. H. (1963). *Methods Carbohydr. Chem.*, **2**, 397.
20. Boons, G.-J., Isles, S. and Setälä, P. (1995). *Synlett*, 755.
21. Zottola, M. A., Alonso, R., Vite, G. D. and Fraser-Reid, B. (1989). *J. Org. Chem.*, **54**, 6123.

22. Lafont, D., Boullanger, P., Cadas, O. and Descotes, G. (1989). *Synthesis*, 191.
23. Rao, M. V. and Nagarajan, M. (1987). *Carbohydr. Res.*, **162**, 141.
24. Caron, S., McDonald, A. I. and Heathcock, C. H. (1996). *Carbohydr. Res.*, **281**, 179.
25. Czernecki, S., Leteux, C. and Veyrières, A. (1992). *Tetrahedron Lett.*, **33**, 221.
26. Mereyala, H. B. and Venkataramanaiah, K. C. (1991). *J. Chem. Res. Synop.*, 197.
27. Haeckel, R., Lauer, G. and Oberdorfer, F. (1996). *Synlett*, 21.
28. Černý, M. (1994). 1,6:2,3- and 1,6:3,4-Dianhydro-β-D-hexopyranoses. Synthesis and preparative applications, in *Frontiers in Biomedicine and Biotechnology. Levoglucosenone and Levoglucosans, Chemistry and Applications*, Witczak, Z. J. ed., vol. 2, ATL Press, Shrewsbury, pp. 121–146.
29. McAuliffe, J. C., Skelton, B. W., Stick, R. V. and White, A. H. (1997). *Aust. J. Chem.*, **50**, 209.
30. Tailler, D., Jacquinet, J.-C., Noirot, A.-M. and Beau, J.-M. (1992). *J. Chem. Soc., Perkin Trans. 1*, 3163.
31. Burgey, C. S., Vollerthun, R. and Fraser-Reid, B. (1994). *Tetrahedron Lett.*, **35**, 2637.
32. Wang, L.-X., Sakairi, N. and Kuzuhara, H. (1990). *J. Chem. Soc., Perkin Trans. 1*, 1677.
33. Lemieux, R. U. and Howard, J. (1963). *Methods Carbohydr. Chem.*, **2**, 400.
34. Lemieux, R. U. and Huber, G. (1956). *J. Am. Chem. Soc.*, **78**, 4117.
35. Halcomb, R. L. and Danishefsky, S. J. (1989). *J. Am. Chem. Soc.*, **111**, 6661.
36. Seeberger, P. H., Bilodeau, M. T. and Danishefsky, S. J. (1997). *Aldrichimica Acta*, **30**, 75.
37. Adam, W., Chan, Y.-Y., Cremer, D., Gauss, J., Scheutzow, D. and Schindler, M. (1987). *J. Org. Chem.*, **52**, 2800.
38. Danishefsky, S. J. and Bilodeau, M. T. (1996). *Angew. Chem. Int. Ed. Engl.*, **35**, 1380.

Deoxy, Amino Deoxy and Branched-chain Sugars

Deoxy, amino deoxy and branched-chain sugars are all characterized by some sort of modification of the carbohydrate superstructure: loss of a hydroxyl group, replacement of a hydroxyl group by an amino group or the addition of an alkyl group. These changes are not without purpose — the resulting molecules often show unique physical characteristics that are associated with some sort of biological role or response.

Deoxy Sugars[1,2]

Deoxy sugars occur commonly in Nature, the most abundant being "deoxy ribose" (2-deoxy-β-D-*erythro*-pentofuranose) as found in deoxyribonucleic acids. Although deoxygenation is possible at any position of a carbohydrate, it is the 6-deoxy sugars that are also widespread — quinovose (6-deoxy-D-glucopyranose), L-rhamnose (6-deoxy-L-mannopyranose) and L-fucose (6-deoxy-L-galactopyranose) are all constituents of molecules that are found, among other places, in plants and bacteria:

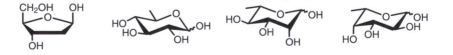

Whereas deoxygenation at C2 in β-D-ribofuranose appears to modify the stability of the resulting nucleic acid, the other three molecules seem to profit from an improved hydrophobic interaction at C6 (CH$_3$) with appropriate receptor sites in biological molecules. Glycals offer a convenient starting point for the synthesis of the less common 2-deoxy sugars:[3]

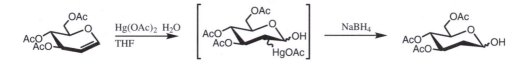

Somewhat rarer but of no less importance are the *dideoxy* sugars that are often found at the terminus of oligosaccharides attached to the surface protein of various bacteria:

paratose
(3,6-dideoxy-D-*ribo*-hexopyranose)

abequose
(3,6-dideoxy-D-*xylo*-hexopyranose)

We are in the fortunate position of being able to appreciate a simple synthesis of paratose:[4]

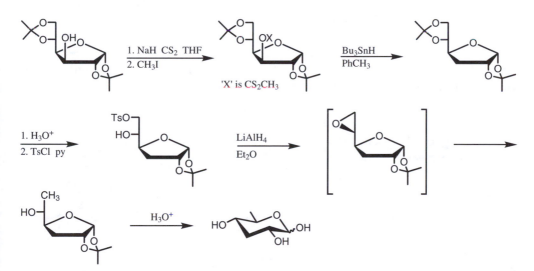

One final point should be made here; with the removal of (at least) one electronegative oxygen atom from a sugar, the so-formed deoxy sugar is more basic than its parent and, as a result, is more prone to protonation. For example, the hydrolysis of methyl 2-deoxy-α-D-*arabino*-hexopyranoside is about 2000 times faster than that of methyl α-D-glucopyranoside.[5]

Amino Deoxy Sugars[6]

Amino deoxy sugars, usually referred to as just the "amino sugars", are no doubt the most important of the modified sugars under discussion here. The most common members are 2-amino- and 2-acetamido-2-deoxy-D-glucopyranose and -D-galactopyranose:

'X' is NH$_2$ or NHAc

Both of these amino sugars are crucial to the well being of many organisms, including humans, because they play pivotal roles in the structure and function of biologically important oligosaccharides, polysaccharides and glycoproteins. The presence of a free amine in a monosaccharide residue allows, of course, for protonation and the generation of a positively charged ammonium ion; the acetamido group, although far less basic than an amine, still imparts a good deal of polarity to the molecule in question.

"*N*-Acetyl-D-glucosamine" is an inexpensive and abundant amino sugar and therefore attracts little synthetic interest as such. However, derivatives of *N*-acetyl-D-glucosamine, much sought after as intermediates in the synthesis of oligosaccharides, are a little more difficult to access:[7]

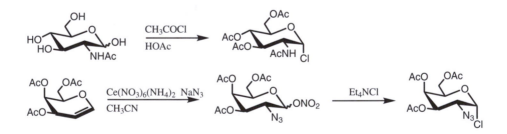

The latter transformation is known as "azidonitration" and, invented by Lemieux, readily provides a masked amino sugar in the form of a glycosyl halide.[8–12]

Whereas *N*-acetyl-D-glucosamine is plentiful and cheap, such is not the case with *N*-acetyl-D-galactosamine. Normally obtained with difficulty from natural sources such as mammalian tissue and cartilage, or as a surrogate from the azidonitration of D-galactal derivatives (above), a simple derivative of *N*-acetyl-D-galactosamine can now be prepared conveniently in a direct sequence:[13]

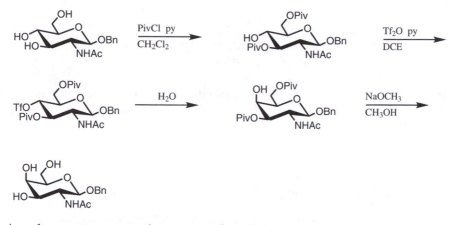

Another common amino sugar found in glycoproteins is *N*-acetylneuraminic acid (5-acetamido-3,5-dideoxy-D-*glycero*-D-*galacto*-2-nonulosonic acid):

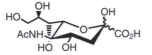

This "higher homologue" amino sugar has received much attention of late, especially in light of its potential to act as the precursor of a drug to combat infection by the influenza virus.[14] Consequently, efficient methods for the large scale production of *N*-acetylneuraminic acid have been devised:[15]

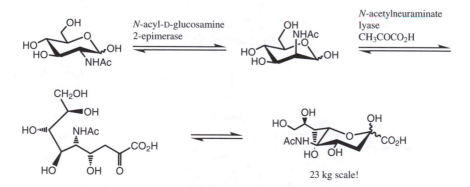

Finally, there exists a class of amino sugar that is better described as a "glycosylamine" — the amino group is attached directly to the anomeric carbon atom:

β-D-glucopyranosylamine

Although such a molecule can be prepared from D-glucose, the product is not particularly stable under aqueous conditions:[16]

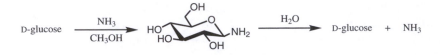

However, with aromatic amines and amino acids, the so-formed glycosylamines are both more stable and more crystalline. When necessary, a glycosyl azide can be a latent precursor of a glycosylamine:[17,18]

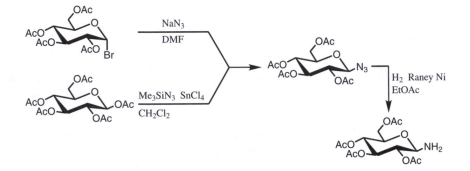

Glycosylamines are implicated as intermediates in the **Amadori rearrangement**, the formation of 1-amino-1-deoxy-2-uloses as by-products when aldoses are treated with amines:[16]

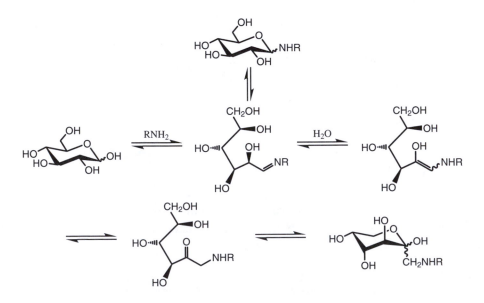

There is a certain similarity here to the Lobry de Bruyn–Alberda van Ekenstein rearrangement discussed earlier and some steps of the Amadori rearrangement are no doubt common to the "Maillard reaction", a complex series of reactions that causes the browning and darkening of sugars when heated in the presence of proteins or their hydrolysis products.[16,19] The "Heyns rearrangement" is a related and useful reaction which converts ketoses into 2-amino-2-deoxy aldoses.[20]

Branched-chain Sugars[21–23]

As the name implies, branched-chain sugars have a carbon substituent at one of the non-terminal carbon atoms of the sugar chain — such substituents may replace either a hydrogen atom or a hydroxyl group (deoxy). Examples of both sorts of branched sugars are found in molecules that often exhibit antibiotic activity:

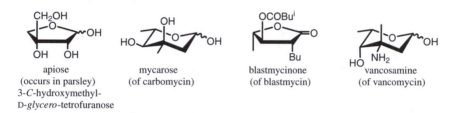

apiose
(occurs in parsley)
3-*C*-hydroxymethyl-
D-*glycero*-tetrofuranose

mycarose
(of carbomycin)

blastmycinone
(of blastmycin)

vancosamine
(of vancomycin)

For the synthesis of both types of branched sugars, uloses can be useful starting materials:

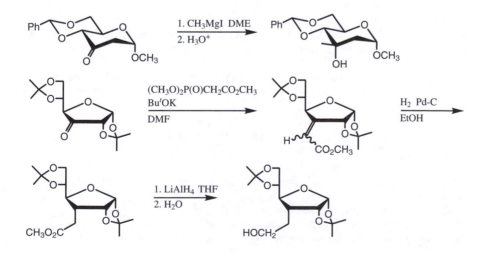

Otherwise, the opening of epoxides and free-radical chemistry can be put to good use:

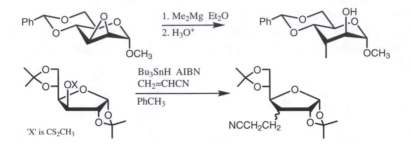

References

1. Hanessian, S. (1966). *Adv. Carbohydr. Chem.*, **21**, 143.
2. Butterworth, R. F. and Hanessian, S. (1971). *Adv. Carbohydr. Chem. Biochem.*, **26**, 279.
3. Bettelli, E., Cherubini, P., D'Andrea, P., Passacantilli, P. and Piancatelli, G. (1998). *Tetrahedron*, **54**, 6011.
4. Patroni, J. J. and Stick, R. V. (1978). *Aust. J. Chem.*, **31**, 445.
5. Capon, B. (1969). *Chem. Rev.*, **69**, 407.
6. Pigman, W. and Horton, D. eds (1980). *The Carbohydrates*, vol. IB. Academic Press, London, p. 643.
7. McLaren, J. M., Stick, R. V. and Webb, S. (1977). *Aust. J. Chem.*, **30**, 2689.
8. Lemieux, R. U. and Ratcliffe, R. M. (1979). *Can. J. Chem.*, **57**, 1244.
9. Kinzy, W. and Schmidt, R. R. (1985). *Liebigs Ann. Chem.*, 1537.
10. Bertozzi, C. R. and Bednarski, M. D. (1992). *Tetrahedron Lett.*, **33**, 3109.
11. Gauffeny, F., Marra, A., Shun, L. K. S., Sinaÿ, P. and Tabeur, C. (1991). *Carbohydr. Res.*, **219**, 237.
12. Seeberger, P. H., Roehrig, S., Schell, P., Wang, Y. and Christ, W. J. (2000). *Carbohydr. Res.*, **328**, 61.
13. Rochepeau-Jobron, L. and Jacquinet, J.-C. (1998). *Carbohydr. Res.*, **305**, 181.
14. von Itzstein, M., Wu, W.-Y., Kok, G. B., Pegg, M. S., Dyason, J. C., Jin, B., Phan, T. V., Smythe, M. L., White, H. F., Oliver, S. W., Colman, P. M., Varghese, J. N., Ryan, D. M., Woods, J. M., Bethell, R. C., Hotham, V. J., Cameron, J. M. and Penn, C. R. (1993). *Nature*, **363**, 418.
15. Maru, I., Ohnishi, J., Ohta, Y. and Tsukada, Y. (1998). *Carbohydr. Res.*, **306**, 575.
16. Pigman, W. and Horton, D., eds (1980). *The Carbohydrates*, vol. IB. Academic Press, London, p. 881.
17. Pfleiderer, W. and Bühler, E. (1966). *Chem. Ber.*, **99**, 3022.
18. Paulsen, H., Györgydeák, Z. and Friedmann, M. (1974). *Chem. Ber.*, **107**, 1568.
19. Ledl, F. and Schleicher, E. (1990). *Ang. Chem. Int. Ed. Engl.*, **29**, 565.
20. Wrodnigg, T. M. and Stütz, A. E. (1999). *Angew. Chem. Int. Ed.*, **38**, 827.
21. Pigman, W. and Horton, D., eds (1980). *The Carbohydrates*, vol. IB. Academic Press, London, p. 761.
22. Yoshimura, J. (1984). *Adv. Carbohydr. Chem. Biochem.*, **42**, 69.
23. Chapleur, Y. and Chrétien, F. (1997). Selected methods for synthesis of branched-chain sugars, in *Preparative Carbohydrate Chemistry*, Hanessian, S. ed., Marcel Dekker, New York, p. 207.

Miscellaneous Reactions

What follows is a selection of chemical reactions that, while important and in common usage in the field of carbohydrates, do not easily fit into any of the sections discussed previously.

Wittig Reaction

The Wittig reaction[1–3] and the closely related Horner–Wadsworth–Emmons variants[4,5] have found much use in carbohydrate synthesis. The two reactions have the potential either to extend the carbohydrate chain (for a hexose, at C1 or C6) or to cause a branching of the chain:[6–9]

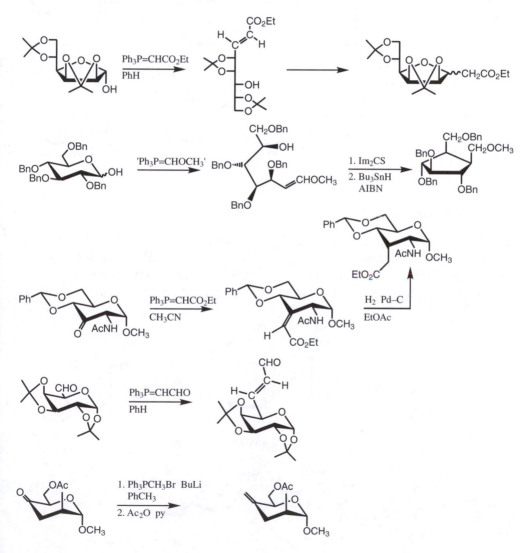

Thiazole-based Homologation[10-12]

The Kiliani–Fischer cyanohydrin synthesis has served early and contemporary chemists equally as well. However, many improvements have been noted and one of the best involves the use of 2-trimethylsilylthiazole for the one-carbon extension of, mainly, aldoses:[13]

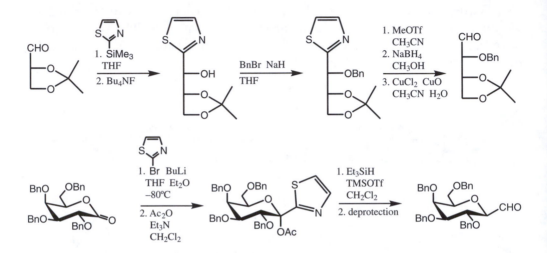

Mitsunobu Reaction[14,15]

The Mitsunobu reaction, involving the activation of a hydroxyl group with the combination of triphenylphosphine and a dialkyl azodicarboxylate and its subsequent replacement with a nucleophile, has been widely used in carbohydrates for the introduction of halogen, oxygen, sulfur and nitrogen:[16-18]

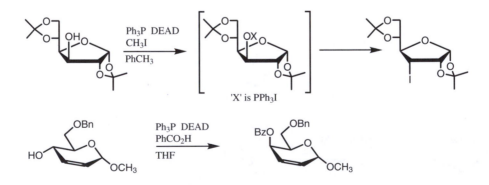

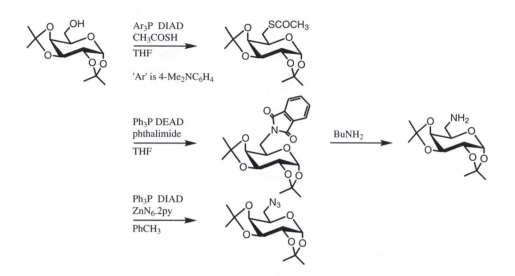

As can be seen from some of the above transformations, the Mitsunobu reaction normally proceeds with inversion of configuration at carbon (an S_N2 process) and this observation fits in nicely with the generally accepted mechanism:[15]

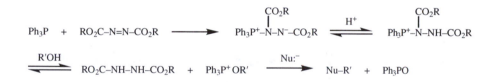

A nice synthesis of a methyl uronate, necessary for the construction of a larger molecule, incorporates many of the reactions discussed here and previously:[19]

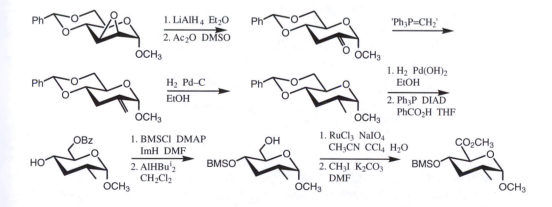

Orthoesters

Orthoesters are encountered not infrequently with carbohydrates and play valuable synthetic roles at both non-anomeric and anomeric positions.

For the synthesis of an orthoester, a diol is generally treated with a simpler orthoester:[20]

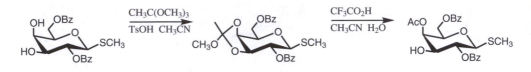

The partial hydrolysis of such an orthoester is a useful reaction, generally leading to a selective protection of the original diol.

For the synthesis of orthoesters at the anomeric position, the conditions devised by Lemieux are in general use:[21,22]

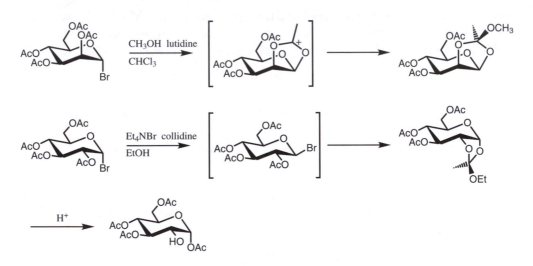

"Amide acetals" can also be used for the synthesis of orthoesters:[23]

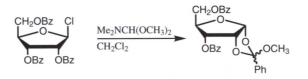

Orthoesters can also be obtained under conditions of kinetic control utilizing 1,1-dimethoxyethene, a reaction somewhat reminiscent of the conversion of diols into acetals utilizing 2-methoxypropene.[24]

A reaction related to orthoester formation, in that it also proceeds through cyclic dioxacarbenium ions, is the transformation of penta-O-acetyl-β-D-glucopyranose into a D-idopyranose salt, a precursor of penta-O-acetyl-α-D-idopyranose:[25]

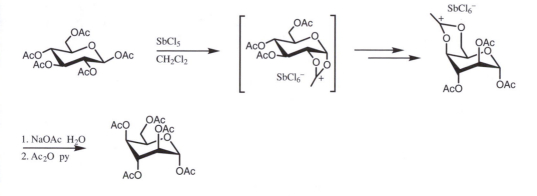

Olefin Metathesis[26,27]

Ring-closing metathesis has become a popular reaction among organic chemists and it came as no surprise to see its immediate application in the synthesis of novel carbohydrate structures:

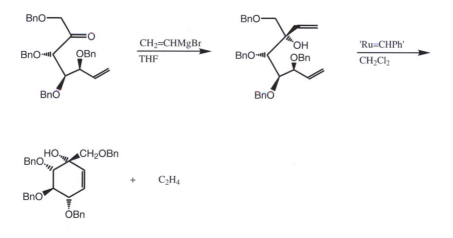

References

1. Zhdanov, Yu. A., Alexeev, Yu. E. and Alexeeva, V. G. (1972). *Adv. Carbohydr. Chem. Biochem.*, **27**, 227.
2. Maryanoff, B. E. and Reitz, A. B. (1989). *Chem. Rev.*, **89**, 863.
3. RajanBabu, T. V. (1997). Functionalized carbocyclic derivatives from carbohydrates: free radical and organometallic methods, in *Preparative Carbohydrate Chemistry*, Hanessian, S. ed., Marcel Dekker, New York, p. 545.

4. Horner, L., Hoffmann, H., Wippel, H. G. and Klahre, G. (1959). *Chem. Ber.*, **92**, 2499.
5. Wadsworth, W. S., Jr. and Emmons, W. D. (1961). *J. Am. Chem. Soc.*, **83**, 1733.
6. Collins, P. M., Overend, W. G. and Shing, T. S. (1982). *J. Chem. Soc., Chem. Commun.*, 297.
7. RajanBabu, T. V. (1991). *Acc. Chem. Res.*, **24**, 139.
8. Brimacombe, J. S., Hanna, R. and Bennett, F. (1985). *Carbohydr. Res.*, **135**, C17.
9. Calvo-Mateo, A., Camarasa, M. J. and de las Heras, F. G. (1984). *J. Carbohydr. Chem.*, **3**, 475.
10. Dondoni, A. and Marra, A. (1997). Thiazole-based one-carbon extension of carbohydrate derivatives, in *Preparative Carbohydrate Chemistry*, Hanessian, S. ed., Marcel Dekker, New York, p. 173.
11. Dondoni, A. (1998). *Synthesis*, 1681.
12. Dondoni, A. and Perrone, D. (1997). *Aldrichimica Acta*, **30**, 35.
13. Dondoni, A., Marra, A. and Perrone, D. (1993). *J. Org. Chem.*, **58**, 275.
14. Mitsunobu, O. (1981). *Synthesis*, 1.
15. Hughes, D. L. (1992). *Organic Reactions*, **42**, 335.
16. Kunz, H. and Schmidt, P. (1982). *Liebigs Ann. Chem.*, 1245.
17. von Itzstein, M., Jenkins, M. J. and Mocerino, M. (1990). *Carbohydr. Res.*, **208**, 287.
18. Viaud, M. C. and Rollin, P. (1990). *Synthesis*, 130.
19. Smith, A. B., III and Hale, K. J. (1989). *Tetrahedron Lett.*, **30**, 1037.
20. Garegg, P. J. and Oscarson, S. (1985). *Carbohydr. Res.*, **136**, 207.
21. Mazurek, M. and Perlin, A. S. (1965). *Can. J. Chem.*, **43**, 1918.
22. Lemieux, R. U. and Morgan, A. R. (1965). *Can. J. Chem.*, **43**, 2199.
23. Hanessian, S. and Banoub, J. (1975). *Carbohydr. Res.*, **44**, C14.
24. Bouchra, M. and Gelas, J. (1998). *Carbohydr. Res.*, **305**, 17.
25. Paulsen, H. (1971). *Adv. Carbohydr. Chem. Biochem.*, **26**, 127.
26. Roy, R. and Das, S. K. (2000). *Chem. Commun.*, 519.
27. Leeuwenburgh, M. A., Appeldoorn, C. C. M., van Hooft, P. A. V., Overkleeft, H. S., van der Marel, G. A. and van Boom, J. H. (2000). *Eur. J. Org. Chem.*, 873.

Chapter 8

The Formation of the Glycosidic Linkage[1-20]

We have reached a watershed in monosaccharides where to discuss any more of their chemistry would be informative, but somewhat trivial and certainly repetitive. It is now time to move on and to consider the broader role of *acetals* in carbohydrate chemistry.

Let us look again at the familiar synthesis of an acetal as performed by Fischer in 1893:

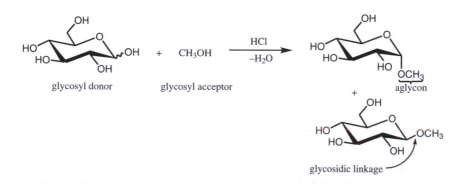

The two products, although certainly acetals, are more commonly referred to as *glycosides* — here, more specifically, methyl α- and β-D-gluco(pyrano)side. The carbohydrate (glycon or glycosyl unit) portion of the molecule is distinguished from the non-carbohydrate *aglycon*. Indeed, the *glycosidic linkage* is formed from a *glycosyl donor* and a *glycosyl acceptor*.

Glycosides are commonly found in the plant and animal kingdoms and in bacteria, for example:

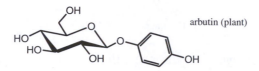

arbutin (plant)

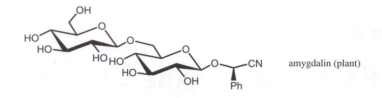

amygdalin (plant)

The latter example, amygdalin, contains in fact **two** glycosidic linkages, one being involved with the aglycon and the other holding two D-glucopyranose units together; enzymatic hydrolysis removes the aglycon and forms the disaccharide, gentiobiose:

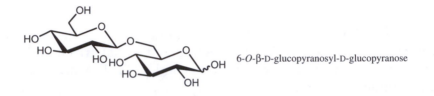

6-*O*-β-D-glucopyranosyl-D-glucopyranose

It is the formation of the glycosidic linkage in molecules such as gentiobiose that is central to the discussion here.

Consider the general formation, now involving a 1,4-linkage, of a disaccharide from two sugars in their pyranose form:

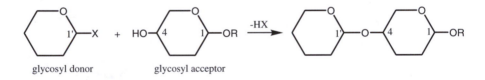

glycosyl donor   glycosyl acceptor

A glycosyl donor, generally of α- or β-configuration, condenses with a glycosyl acceptor (elimination of HX) to form the disaccharide containing the new glycosidic linkage, ideally of the α- or β-configuration at C1′. In the process, there is *no change in configuration* at C4 in the glycosyl acceptor.

In less common circumstances, the glycosyl acceptor may be the hydroxyl group of the anomeric (hemiacetal) centre:

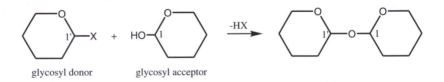

glycosyl donor   glycosyl acceptor

The formation of the glycosidic linkages (there are now two!) results in the α- or β-configuration at both C1 and C1′ — the product is a non-reducing disaccharide. Such a disaccharide is trehalose:

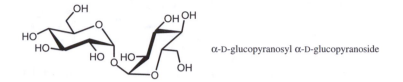

α-D-glucopyranosyl α-D-glucopyranoside

It will be apparent, even at this stage, that the formation of a glycosidic linkage will not be an easy task. Apart from the activation of the glycosyl donor, there are the problems of the stereoselectivity (α- or β-) of the process and access to just the desired hydroxyl group of the glycosyl acceptor (protecting group chemistry). Nature, of course, circumvents all of these problems with the use of enzymes but, for the synthetic carbohydrate chemist, much ingenuity, creativity and hard work are necessary to match the rewards of evolution!

The Different Glycosidic Linkages

In forming the glycosidic linkage, close attention must be paid to the orientation of the hydroxyl group at C2 of a pyranose ring — there are four common outcomes:

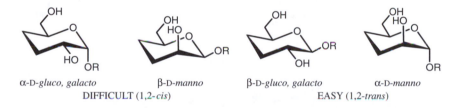

α-D-*gluco, galacto*   β-D-*manno*   β-D-*gluco, galacto*   α-D-*manno*
DIFFICULT (1,2-*cis*)   EASY (1,2-*trans*)

As will be seen, it is easy to form 1,2-*trans* linkages but harder to form the 1,2-*cis* linkage. In addition, the hydroxyl group at C2 must be protected during the actual glycosidation step (otherwise unwanted intra- or intermolecular reactions could occur) and the nature of this protecting group can have a profound effect on the stereochemistry of the newly introduced aglycon at C1. The whole situation changes when there is no substituent or an amide group at C2 (a 2-deoxy sugar and a 2-acetamido-2-deoxy sugar, respectively); mention will be made of these exceptions later.

General Comments on the Formation of the Glycosidic Linkage

Nearly all the methods available for the formation of the glycosidic linkage utilize a glycosyl donor that is a precursor of either an intermediate oxacarbenium ion or, at least, a species that has significant positive charge at the anomeric carbon atom:

As such, these "oxacarbenium ions" are subject to all the normal factors that stabilize/destabilize such short-lived, high energy species and it is appropriate to mention some of them here. Some of these factors also affect the nucleophilicity and, hence, the reactivity of the glycosyl acceptor.

Ion-pairs and Solvent

The ionization of a glycosyl donor at the anomeric carbon generates a salt which, depending on the solvent, can have characteristics ranging from an intimate (tight) ion-pair to a solvent-separated ion-pair. The anion of the ion-pair may shield one face of the oxacarbenium ion from the approach of the glycosyl acceptor or, if anion exchange can occur, it may be the opposite face that is shielded:

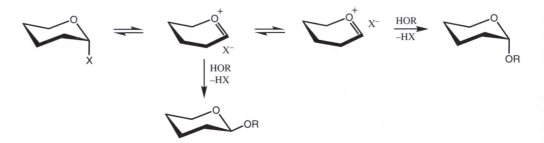

The role of the solvent may be passive, or active. Solvents of high dielectric constant obviously can stabilize a positive charge very well; those solvents with basic lone pairs of electrons (ethyl ether, tetrahydrofuran) can do so, as well, and reversibly.[21,22] One solvent which has the dual characteristics of moderate polarity and basicity is acetonitrile; the results of many glycosidation reactions performed in acetonitrile can only be explained by solvent intervention:

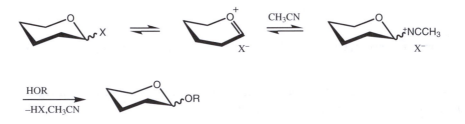

The Substituent at C2

Because the substituent at C2 is a functionalized (protected) hydroxyl group, the choices are generally limited to an ether or an ester, the common groups being a benzyl ether (4-methoxybenzyl and silyl ethers are not uncommon) and an acetic, benzoic or pivalic (2,2-dimethylpropanoic) acid ester. The role of the benzyl ether is unique — it is an inert group that provides a degree of steric hindrance to any incoming nucleophile (the glycosyl acceptor). The ester, on the other hand, not only provides a degree of steric encumbrance (especially for that of the pivalic acid ester) but also offers the possibility of chemical intervention to assist in the formation of the glycoside or, in other cases, an orthoester:

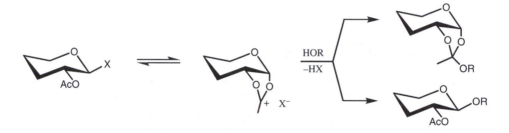

We have encountered orthoesters before and will return to the subject soon.

The "Armed/Disarmed" Concept

It is an experimental fact that tetra-*O*-acetyl-α-D-glucopyranosyl bromide is a white, crystalline solid that can be stored in cold, dry conditions for several months. On the other hand, tetra-*O*-benzyl-α-D-glucopyranosyl bromide is an unstable compound that is never isolated and purified, but simply used as generated.

It was Paulsen[3] who first noted these differences in stability but some years had to pass before Fraser-Reid, in his incisive and wide-ranging studies on 4-pentenyl glycosides, formalized the result with the "armed/disarmed" concept.[23] Put simply, the benzyl ethers are electronically passive groups that do not discourage the development of positive charge in the pyranose ring, most commonly at C1. Conversely, the acetate groups (especially when located at C2) are electron withdrawing and so disfavour any build up of positive charge in the ring (C1). The result is "arming" of the benzylated bromide and "disarming" of the ester counterpart. Other functional groups show similar effects.[24]

The "armed/disarmed" concept can also be applied, although to a much lesser degree, to the glycosyl acceptor in that electronically passive (conversely, electron withdrawing) groups will little affect (decrease) the nucleophilicity of the hydroxyl group in question and so make for a faster (slower) glycosidation.

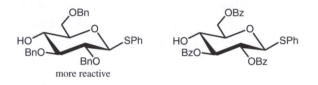

more reactive

The "Torsional Control" Concept

Again, it was Fraser-Reid who noted that the presence of another ring fused to the pyranose ring could have an influence on events occurring at the anomeric centre. For example, the presence of a 4,6-*O*-benzylidene ring hinders the necessary flattening of the pyranose ring that accompanies oxacarbenium ion formation and so slows the process of glycosylation:[25]

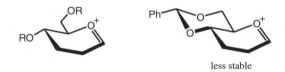

less stable

This "torsional control" concept was very timely for Ley in his recent work on the protection of *trans* vicinal diols with diacetal reagents. Such diacetals again impart a degree of rigidity on the pyranose ring and discourage the ready formation of oxacarbenium ions; Ley has introduced the term "reactivity tuning" to describe the process:[26–28]

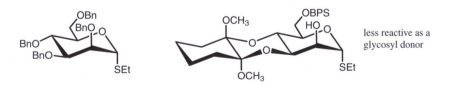

less reactive as a glycosyl donor

The "Latent/Active" Concept

Some "latent" glycosyl donors are characterized by having a group at the anomeric carbon that is stable under most of the conditions employed in glycosidation reactions, yet can be manipulated later in a synthetic sequence to provide an "active" donor.[29] For example, but-3-en-2-yl 3,4,6-tri-*O*-benzyl-β-D-glucoside can act only as a glycosyl acceptor, whereas the related but-2-en-2-yl β-D-glucoside (obtained by acetylation and isomerization with Wilkinson's catalyst) is an active glycosyl donor owing to the presence of a reactive enol ether:[30]

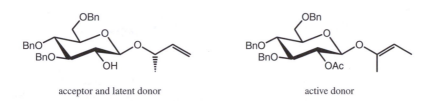

acceptor and latent donor   active donor

Activation of the Glycosyl Acceptor

There is no point in having a good glycosyl donor if the acceptor is too unreactive for glycosidation to occur. In order to increase the reactivity of some acceptors, and to impart a degree of chemoselectivity to others, various modifications have been made to the hydroxyl group(s) of the acceptor. For example, trityl,[31] silyl[32] and stannyl[33] ethers all seem to polarize the bonding electrons towards the oxygen atom, the first because of the stability of any positive charge that develops on carbon and the other two because of the electropositive nature of silicon and of tin. Whatever the reason, the consequence is a more nucleophilic oxygen atom in the acceptor:

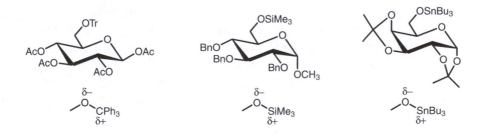

The Concept of "Orthogonality"

The concept of "orthogonality" was formalized by Tomoya Ogawa and rapidly embraced by the carbohydrate community.[34] In essence, as applied to glycosyl donors, orthogonality requires two separate donors to have different groups at the anomeric carbon, but each activatable in a unique and discrete manner. In its simplest form, orthogonality allows for the production of different glycosidic linkages:

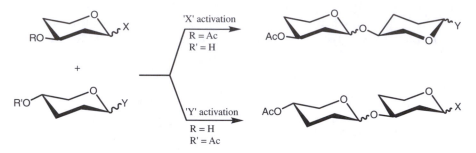

The concept has been so successful that it is also applied to the various protecting groups found on a pyranose ring; the thioglycoside below is protected in a fully orthogonal manner, with each group being removed by a different set of reaction conditions:[35]

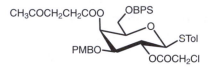

"Double stereodifferentiation", or "double asymmetric induction", involving "matched" and "mismatched" pairs, may occasionally have a significant effect on the stereoselectivity of glycosidation.[36–38]

Before we proceed, a final few words on the **Fischer glycosidation**. The strength of this method is its simplicity — a monosaccharide is placed in a large volume of the appropriate alcohol, a small amount of acetyl chloride is added (to generate hydrogen chloride) and the mixture is heated at reflux.[39] Depending on the nature of the monosaccharide and of the desired outcome, reaction times may be short (to generate products of kinetic control, for example, furanosides) or long (products of thermodynamic control, pyranosides). The main limitation to the process is that, because of the low activity of the glycosyl donor employed (a free sugar or hemiacetal), only reactive alcohols (glycosyl acceptors) can be employed. Some examples are given below, together with a very simple mechanistic rationalization:[40, 41]

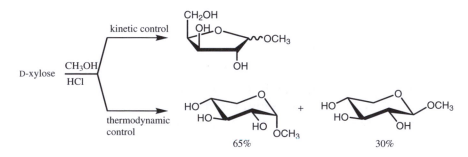

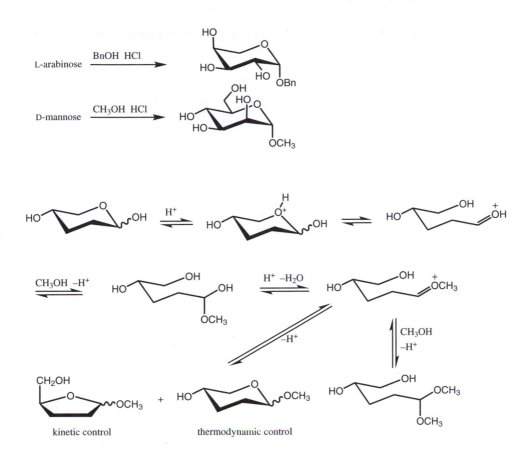

Related to the Fischer glycosidation, in that the glycosyl "donor" is also the somewhat unreactive free sugar, is the synthesis of glycosides by the direct alkylation of the hemiacetal hydroxyl group:

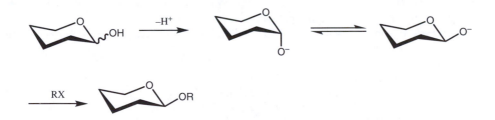

Under the strongly basic conditions employed, the free sugar equilibrates with a mixture of the anions derived by deprotonation of the anomeric hydroxyl group; preferential alkylation of the oxyanion derived from the β-anomer of the free sugar leads directly to the β-D-glycoside. This method of glycosidation is a rare example in which the oxygen atom of the glycosyl "donor" is retained and there is ample opportunity for inversion of configuration at carbon in the alkylating agent. Some examples of successful glycosidations follow:[42–45]

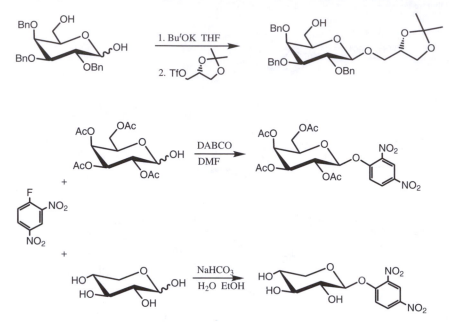

This now brings us to the stage where we can discuss the various methods that are used to form the glycosidic linkage. The "big three" glycosyl donors (halides, trichloroacetimidates and thioglycosides) will be discussed first, followed by a handful of less common (but still important) luminaries. Finally, a listing will be given of the less important or "still rising" donors. Where possible, attention to mechanistic rationalization will be given.

References

1. Overend, W. G. (1972). Glycosides, in *The Carbohydrates*, Pigman, W. and Horton, D. eds, vol. IA, Academic Press, London, p. 279.
2. Bochkov, A. F. and Zaikov, G. E. (1979). *Chemistry of the O-Glycosidic Bond: Formation and Cleavage*, Pergamon Press, Oxford.
3. Paulsen, H. (1982). *Angew. Chem. Int. Ed. Engl.*, **21**, 155.
4. Paulsen, H. (1984). *Chem. Soc. Rev.*, **13**, 15.
5. Schmidt, R. R. (1986). *Angew. Chem. Int. Ed. Engl.*, **25**, 212.
6. Schmidt, R. R. (1989). *Pure Appl. Chem.*, **61**, 1257.
7. Sinaÿ, P. (1991). *Pure Appl. Chem.*, **63**, 519.
8. Schmidt, R. R. (1991). Synthesis of glycosides, in *Comprehensive Organic Synthesis*, Trost, B. M. and Fleming, I. eds, vol. 6, Pergamon, Oxford, p. 33.
9. Lockhoff, O. (1992). Acetale als anomere zentren von kohlenhydraten (hal/O- und O/O-acetale), in *Methoden der organischen chemie* (Houben-Weyl) E14a/3, Thieme, Stuttgart, p. 621.

10. Darcy, R. and McCarthy, K. (1993). Disaccharides and oligosaccharides, in *Rodd's Chemistry of Carbon Compounds*, Sainsbury, M. ed., 2nd edn, 2nd supplements, vol. IE/F/G. Elsevier, Amsterdam, p. 437.
11. Toshima, K. and Tatsuta, K. (1993). *Chem. Rev.*, **93**, 1503.
12. Barresi, F. and Hindsgaul, O. (1995). Glycosylation methods in oligosaccharide synthesis, in *Modern Synthetic Methods*, Ernst, B. and Leumann, C. eds, Verlag Helvetica Chimica Acta, Basel, p. 281.
13. Boons, G.-J. (1996). *Contemp. Org. Synth.*, **3**, 173.
14. Boons, G.-J. (1996). *Tetrahedron*, **52**, 1095.
15. Boons, G.-J. (1996). *Drug Discovery Today*, **1**, 331.
16. Paulsen, H. (1996). Twenty five years of carbohydrate chemistry; an overview of oligosaccharide synthesis, in *Modern Methods in Carbohydrate Synthesis*, Khan, S. H. and O'Neill, R. A. eds, Harwood Academic, Netherlands, p. 1.
17. von Rybinski, W. and Hill, K. (1998). *Angew. Chem. Int. Ed.*, **37**, 1328.
18. Whitfield, D. M. and Douglas, S. P. (1996). *Glycoconjugate J.*, **13**, 5.
19. Schmidt, R. R., Castro-Palomino, J.-C. and Retz, O. (1999). *Pure Appl. Chem.*, **71**, 729.
20. Davis, B. G. (2000). *J. Chem. Soc., Perkin Trans. 1*, 2137.
21. Wulff, G., Schröder, U. and Wichelhaus, J. (1979). *Carbohydr. Res.*, **72**, 280.
22. Jünnemann, J., Lundt, I. and Thiem, J. (1991). *Liebigs Ann. Chem.*, 759.
23. Fraser-Reid, B., Wu, Z., Udodong, U. E. and Ottosson, H. (1990). *J. Org. Chem.*, **55**, 6068.
24. Zhang, Z., Ollmann, I. R., Ye, X.-S., Wischnat, R., Baasov, T. and Wong, C.-H. (1999). *J. Am. Chem. Soc.*, **121**, 734.
25. Fraser-Reid, B., Wu, Z., Andrews, C. W., Skowronski, E. and Bowen, J. P. (1991). *J. Am. Chem. Soc.*, **113**, 1434.
26. Douglas, N. L., Ley, S. V., Lücking, U. and Warriner, S. L. (1998). *J. Chem. Soc., Perkin Trans. 1*, 51.
27. Green, L., Hinzen, B., Ince, S. J., Langer, P., Ley, S. V. and Warriner, S. L. (1998). *Synlett*, 440.
28. Grice, P., Ley, S. V., Pietruszka, J., Priepke, H. W. M. and Walther, E. P. E. (1995). *Synlett*, 781.
29. Cao, S., Hernández-Matéo, F. and Roy, R. (1998). *J. Carbohydr. Chem.*, **17**, 609.
30. Boons, G.-J. and Isles, S. (1994). *Tetrahedron Lett.*, **35**, 3593.
31. Tsvetkov, Y. E., Kitov, P. I., Backinowsky, L. V. and Kochetkov, N. K. (1996). *J. Carbohydr. Chem.*, **15**, 1027.
32. Ziegler, T. (1998). *J. prakt. Chem.*, **340**, 204.
33. Ogawa, T. and Matsui, M. (1976). *Carbohydr. Res.*, **51**, C13.
34. Kanie, O., Ito, Y. and Ogawa, T. (1994). *J. Am. Chem. Soc.*, **116**, 12073.
35. Wong, C.-H., Ye, X.-S. and Zhang, Z. (1998). *J. Am. Chem. Soc.*, **120**, 7137.
36. Masamune, S., Choy, W., Petersen, J. S. and Sita, L. R. (1985). *Angew. Chem. Int. Ed. Engl.*, **24**, 1.
37. Spijker, N. M. and van Boeckel, C. A. A. (1991). *Angew. Chem. Int. Ed. Engl.*, **30**, 180.
38. Ziegler, T. and Lemanski, G. (1998). *Eur. J. Org. Chem.*, 163.
39. Mowery, D. F., Jr. (1963). *Methods Carbohydr. Chem.*, **2**, 328.
40. Capon, B. (1969). *Chem. Rev.*, **69**, 407.
41. Garegg, P. J., Johansson, K.-J., Konradsson, P. and Lindberg, B. (1999). *J. Carbohydr. Chem.*, **18**, 31.
42. Schmidt, R. R. (1996). The anomeric *O*-alkylation and the trichloroacetimidate method — versatile strategies for glycoside bond formation, in *Modern Methods in Carbohydrate Synthesis*, Khan, S. H. and O'Neill, R. A. eds, Harwood Academic, Netherlands, p. 20.

43. Schmidt, R. R. and Kinzy, W. (1994). *Adv. Carbohydr. Chem. Biochem.*, **50**, 21.
44. Koeners, H. J., de Kok, A. J., Romers, C. and van Boom, J. H. (1980). *Recl. Trav. Chim. Pays-Bas*, **99**, 355.
45. Sharma, S. K., Corrales, G. and Penadés, S. (1995). *Tetrahedron Lett.*, **36**, 5627.

Glycosyl Halides

Glycosyl halides play a historical role in the development of the glycosidic linkage in that it was a glycosyl chloride that was used in the first synthesis (1879) of a glycoside:[1]

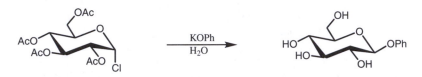

Some years later (1901), Koenigs and Knorr[a,b] extended the approach by treating "acetobromglucose" with alcohols in the presence of silver(I) carbonate:[2]

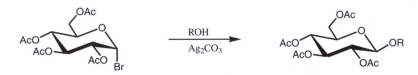

Since these two seminal announcements, glycosyl halides have been at the forefront of new methodology for the synthesis of the glycosidic linkage. Glycosyl chlorides and bromides are used routinely in glycoside synthesis; glycosyl iodides, for long considered to be too unstable to be of any use, are now finding their rightful place; glycosyl fluorides, the most stable of the glycosyl halides, have broadened the whole approach towards the formation of the glycosidic linkage.

What follows now will be a "cameo" discussion of the various methods that utilize glycosyl halides, with some attention to mechanistic rationale.

The Koenigs–Knorr Reaction[4,5] (1,2-*trans*)

Since the inception of the Koenigs–Knorr reaction, there has been an enormous effort to improve the process by utilizing a co-solvent (for acceptors more

---

[a] William Koenigs, or Königs (1851–1906), student of von Baeyer and fellow student with Emil Fischer. Edward Knorr, student of Koenigs.

[b] It is not a generally known fact that Fischer and Armstrong published very similar findings just months after the paper by Koenigs and Knorr.[3]

complex than simple alcohols), adding a desiccant (to absorb any liberated water),[6] adding powdered molecular sieves (to absorb both the liberated hydrogen halide and adventitious water), adding a trace of elemental iodine (to suppress side reactions?)[7] and distinguishing the roles of "promoter" versus "acid acceptor". In this last regard, other heavy metal salts have been of great use — the combination of mercury(II) bromide/mercury(II) oxide seems to offer the advantages of promotion (HgBr$_2$, a synergistic effect?) and acid acceptor (HgO).[8] Soluble promoters, such as mercury(II) cyanide, used in acetonitrile or nitromethane and introduced by Helferich, and silver(I) triflate, are commonly used in the Koenigs–Knorr reaction.[9,10] Some examples of successful glycosidations under Koenigs–Knorr conditions follow:

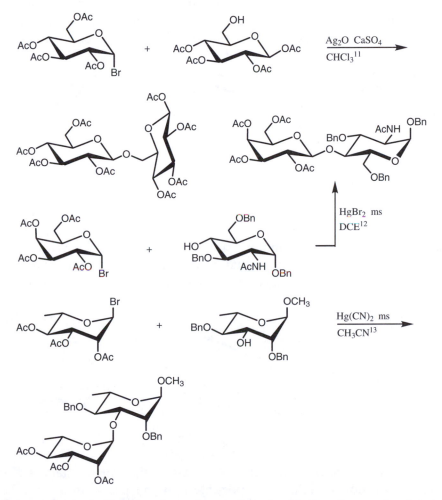

A great deal of effort has been devoted to elucidating the mechanism of the Koenigs–Knorr reaction. For the heterogeneous process [insoluble silver(I) and mercury(II) salts], it appears that a "bimolecular" process operates, sometimes described as a "push-pull" mechanism, resulting in inversion of stereochemistry at the anomeric centre:

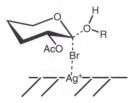

For the homogeneous process [soluble silver(I) and mercury(II) compounds], it is suggested that a "unimolecular" process operates and heterolysis of the C1-halogen bond results in an anomeric carbocation, stabilized by resonance and undergoing a subsequent reaction with the neighbouring ester group at C2:

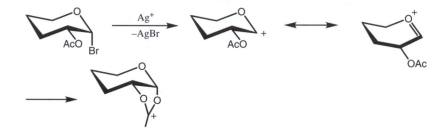

The fate of the new cyclic carbocation is now twofold: approach of the alcohol at the anomeric carbon leads directly to the *trans*-glycoside, whereas attack at the actual carbocation results in the formation of an orthoester. With the acidic conditions (both protic and Lewis) that prevail in a Koenigs–Knorr reaction, a rearrangement of any preformed orthoester into the observed glycoside is probable:[14]

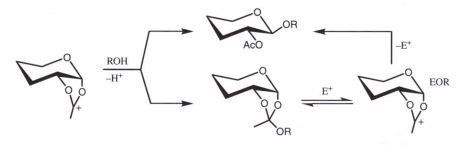

In light of recent studies and the observation that orthoesters are often significant by-products, it is quite possible that the preparation of 1,2-*trans* glycosides using the Koenigs–Knorr reaction proceeds *via* an orthoester intermediate.[15,16]

The use of the Koenigs–Knorr reaction for the large-scale preparation of glycosides is not recommended, owing to the accumulation of toxic, heavy metal wastes. Although other promotors of the reaction are available,[17] their use is limited; this has resulted in the evolution, and refinement, of other methods for the synthesis of 1,2-*trans* glycosides.

The Orthoester Procedure[4,18] (1,2-*trans*)

As we have seen on several occasions, simple orthoesters result most conveniently from the treatment of a glycosyl halide with an alcohol in the presence of a base and, where necessary, a source of halide ions. These simple orthoesters can give rise to more complex orthoesters by a process of acid-catalysed transesterification; during this process, distillation or the addition of the appropriate molecular sieve usually removes the volatile alcohol. The subsequent rearrangement of the more complex orthoesters into 1,2-*trans* glycosides was initially performed with Lewis acids such as mercury(II) bromide; more recently, trimethylsilyl triflate has appeared as the reagent of choice:[19]

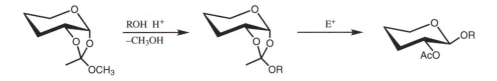

On occasion, especially with orthoacetates, a major and unwanted by-product is the acetic acid ester of the acceptor alcohol; this complication can essentially be eliminated by the use of orthobenzoates:[20]

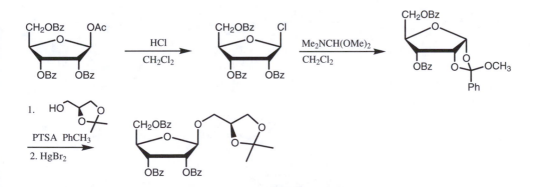

Kochetkov has made the orthoester approach virtually his own. In a lifetime of publications, subtle changes such as the use of *tert*-butyl orthoesters[21] or more major developments involving the introduction of cyano- or thio-ethylidene analogues[22] have left his name indelibly stamped on the method. In this last regard, the ideal orthoester procedure involves a trityl ether as acceptor and trityl perchlorate as catalyst:

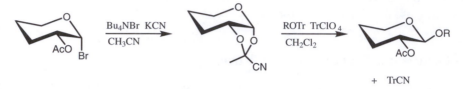

An interesting extension of Kochetkov's work involves the preparation of a tritylated cyanoethylidene derivative of α-D-glucopyranose, capable of polymerization in the presence of trityl perchlorate:

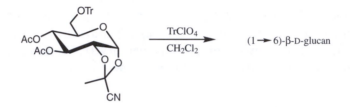

The mechanisms of the orthoester procedures have been thoroughly studied and the various outcomes rationalized:

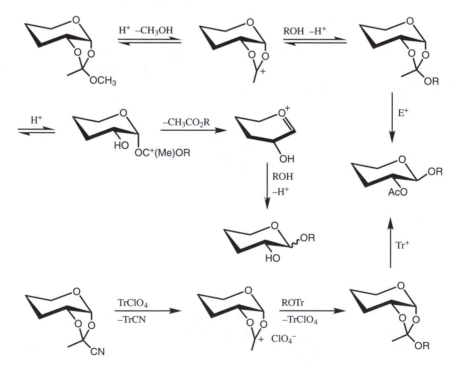

A very recent synthesis of disaccharides involving a $(1 \rightarrow 4)$-β-D linkage utilizes the stereoselective reduction of an anomeric orthoester:[23]

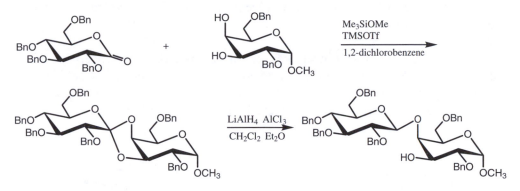

Halide Catalysis (1,2-*cis*)

It is a credit to the ingenuity and creativity of Lemieux that, some twenty-five years ago, he was solely responsible for the development of a method for the synthesis of 1,2-*cis* glycosides, termed "halide catalysis", that is still in common use today.[24] Lemieux, based on an analysis of the anomerization and subsequent reactions of a variety of glycosyl halides, reasoned that a 2-*O*-benzyl protected α-D-glycosyl bromide could be rapidly equilibrated with the β-D-anomer by catalysis involving a tetraalkylammonium bromide. This highly reactive β-D-anomer, or more probably an ion-pair derived from it, then proceeds on to the observed product, the α-D-glycoside. The whole process relies on the facial availability of various ion-pairs and subtle changes in the conformation of the many intermediates:

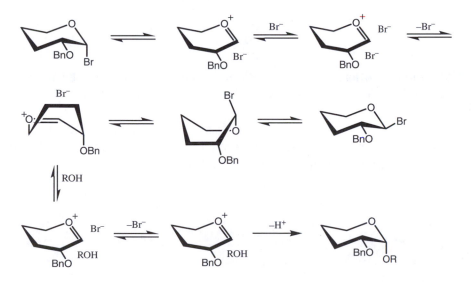

In view of the complexity (and beauty) of the process, it is not surprising that Lemieux took umbrage with those who describe the method as "*in situ* anomerization".[25] Some recent mechanistic studies have indicated that the "halide catalysis" procedure is first order in both the donor and the acceptor.[26]

Many 1,2-*cis* glycosides have been prepared since the announcement of the method in 1975 — α-D-glucopyranosides and α-D-galactopyranosides are examples of such linkages, with α-L-fucopyranosides, a common component of many naturally occurring oligosaccharides, being a special bonus:

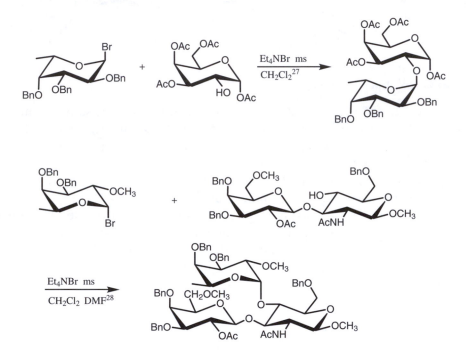

It was found advantageous to add both molecular sieves (to absorb the liberated hydrogen bromide) and dimethylformamide (an apparent catalyst for the reaction).[27-30]

In the early years of the "halide catalysis" procedure, the glycosyl bromides were generated from the treatment of a 4-nitrobenzoate with hydrogen bromide; nowadays, more convenient procedures are available:[24, 31]

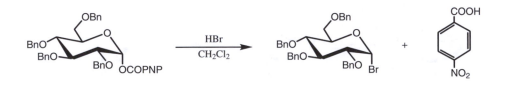

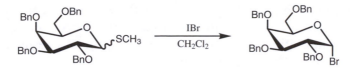

Gervay has recently reported methods for the synthesis of α-D-glycosyl iodides and has found them to be "superior substrates" for the "halide catalysis" protocol.[32]

Glycosyl Fluorides[33–35] (1,2-*cis* and 1,2-*trans*)

So far, we have seen that the relatively stable glycosyl chlorides have been of limited use in glycoside synthesis. On the other hand, glycosyl bromides show a versatility that allows their use in the Koenigs–Knorr, orthoester and "halide catalysis" procedures. Glycosyl iodides, of limited stability, show some promise for an improvement to the "halide catalysis" process. It seems that the outcast of the glycosyl halides is the glycosyl fluoride and this was to remain so until about 1980. Before then, methods of preparation were limited, even β-D-glycosyl fluorides were relatively stable (and the α-D-anomer even more so!), and there existed no "promoter" to encourage a glycosyl fluoride to act as a donor.

Mukaiyama changed things[36] with the introduction of a range of (exotic) promoters to cause glycosyl fluorides to act as respectable glycosyl donors:[37]

$SnCl_2$, $AgClO_4$	Et_2OBF_3
$SnCl_2$, $TrClO_4$	SiF_4
Cp_2MCl_2 (M = Ti, Zr, Hf)	TiF_4
$AgClO_4$	Me_3SiOTf
Me_2GaCl	Tf_2O
$La(ClO_4)_3 \cdot nH_2O$	$Yb(OTf)_3$
$TrB(C_6F_5)_4$	

Nicolaou became involved in this renaissance of glycosyl fluorides when he showed that a rather benign thioacetal (thioglycoside) could be converted into a fluoride for subsequent use as a glycosyl donor:[38]

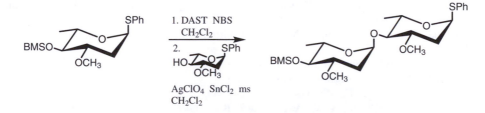

In general, glycosyl fluorides can be used for the synthesis of both 1,2-*cis* and 1,2-*trans* glycosides. For donors having a participating (ester) group at C2, the outcome usually follows the principles of the Koenigs–Knorr reaction and a 1,2-*trans* glycoside results; with a non-participating (ether) group at C2, inversion of configuration at the anomeric carbon generally results, presumably the consequence of a bimolecular process:

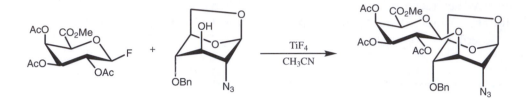

Again, the choice of solvent can play a critical role in the outcome of a particular glycosidation.[39] Ley has recently described the application of 1,2-diacetal protecting groups to control the "reactivity tuning" of glycosyl fluorides in glycoside synthesis.[40]

References

1. Michael, A. (1879, 1885). *Am. Chem. J.*, **1**, 305; **6**, 336.
2. Koenigs, W. and Knorr, E. (1901). *Ber. Dtsch. Chem. Ges.*, **34**, 957.
3. Fischer, E. and Armstrong, E. F. (1901). *Ber. Dtsch. Chem. Ges.*, **34**, 2885.
4. Bochkov, A. F. and Zaikov, G. E. (1979). *Chemistry of the O-Glycosidic Bond: Formation and Cleavage*, Pergamon Press, Oxford.
5. Igarashi, K. (1977). *Adv. Carbohydr. Chem. Biochem.*, **34**, 243.
6. Helferich, B. and Gootz, R. (1931). *Ber. Dtsch. Chem. Ges.*, **64**, 109.
7. Helferich, B., Bohn, E. and Winkler, S. (1930). *Ber. Dtsch. Chem. Ges.*, **63**, 989.
8. Flowers, H. M. (1972). *Methods Carbohydr. Chem.*, **6**, 474.
9. Helferich, B. and Zirner, J. (1962). *Chem. Ber.*, **95**, 2604.
10. Hanessian, S. and Banoub, J. (1980). *Methods Carbohydr. Chem.*, **8**, 247.
11. Reynolds, D. D. and Evans, W. L. (1938). *J. Am. Chem. Soc.*, **60**, 2559.
12. Jacquinet, J.-C., Duchet, D., Milat, M.-L. and Sinaÿ, P. (1981). *J. Chem. Soc., Perkin Trans. 1*, 326.
13. Lipták, A., Nánási, P., Neszmélyi, A. and Wagner, H. (1980). *Tetrahedron*, **36**, 1261.
14. Crich, D., Dai, Z. and Gastaldi, S. (1999). *J. Org. Chem.*, **64**, 5224.
15. Garegg, P. J., Konradsson, P., Kvarnström, I., Norberg, T., Svennson, S. C. T. and Wigilius, B. (1985). *Acta Chem. Scand.*, **B39**, 569.
16. Nukada, T., Berces, A., Zgierski, M. Z. and Whitfield, D. M. (1998). *J. Am. Chem. Soc.*, **120**, 13291.
17. Kartha, K. P. R., Aloui, M. and Field, R. A. (1996). *Tetrahedron Lett.*, **37**, 8807.
18. Pacsu, E. (1945). *Adv. Carbohydr. Chem.*, **1**, 77.
19. Wang, W. and Kong, F. (1998). *J. Org. Chem.*, **63**, 5744.
20. McAdam, D. P., Perera, A. M. A. and Stick, R. V. (1987). *Aust. J. Chem.*, **40**, 1901.
21. Kochetkov, N. K., Bochkov, A. F., Sokolovskaya, T. A. and Snyatkova, V. J. (1971). *Carbohydr. Res.*, **16**, 17.
22. Kochetkov, N. K. (1987). *Tetrahedron*, **43**, 2389.

23. Ohtake, H., Iimori, T. and Ikegami, S. (1998). *Synlett*, 1420.
24. Lemieux, R. U., Hendriks, K. B., Stick, R. V. and James, K. (1975). *J. Am. Chem. Soc.*, **97**, 4056.
25. Khan, S. H. and O'Neill, R. A., eds (1996). *Modern Methods in Carbohydrate Synthesis*, Harwood Academic, Netherlands, p. xi.
26. Bowden, T., Garegg, P. J., Konradsson, P. and Maloisel, J.-L., unpublished results.
27. Lemieux, R. U. and Driguez, H. (1975). *J. Am. Chem. Soc.*, **97**, 4069.
28. Nikrad, P. V., Beierbeck, H. and Lemieux, R. U. (1992). *Can. J. Chem.*, **70**, 241.
29. Lemieux, R. U. and Driguez, H. (1975). *J. Am. Chem. Soc.*, **97**, 4063.
30. Hindsgaul, O., Norberg, T., Le Pendu, J. and Lemieux, R. U. (1982). *Carbohydr. Res.*, **109**, 109.
31. Kartha, K. P. R. and Field, R. A. (1997). *Tetrahedron Lett.*, **38**, 8233.
32. Gervay, J. (1998). Glycosyl iodides in organic synthesis, in *Organic Synthesis: Theory and Applications*, Hudlicky, T. ed., vol. 4, JAI Press, Greenwich CT, p. 121.
33. Dax, K., Albert, M., Ortner, J. and Paul, B. J. (1999). *Current Org. Chem.*, **3**, 287.
34. Tsuchiya, T. (1990). *Adv. Carbohydr. Chem. Biochem.*, **48**, 91.
35. Thiem, J. and Wiesner, M. (1993). *Carbohydr. Res.*, **249**, 197.
36. Mukaiyama, T., Murai, Y. and Shoda, S.-i. (1981). *Chem. Lett.*, 431.
37. Takeuchi, K. and Mukaiyama, T. (1998). *Chem. Lett.*, 555.
38. Nicolaou, K. C., Dolle, R. E., Papahatjis, D. P. and Randall, J. L. (1984). *J. Am. Chem. Soc.*, **106**, 4189.
39. Kreuzer, M. and Thiem, J. (1986). *Carbohydr. Res.*, **149**, 347.
40. Baeschlin, D. K., Green, L. G., Hahn, M. G., Hinzen, B., Ince, S. J. and Ley, S. V. (2000). *Tetrahedron*: Asymmetry, **11**, 173.

The Trichloroacetimidate Method[1,2] (1,2-*cis* and 1,2-*trans*)

Sinaÿ was the first to use an imidate for the synthesis of a glycoside,[3] but it was to be Schmidt who would extend the method and make the use of trichloroacetimidates a rival to the well established Koenigs–Knorr procedure. In fact, the trichloroacetimidate (TCA) method is now often preferred for the synthesis of a 1,2-*trans*-glycoside, simply because it does not involve the use of heavy metal reagents in the promotion step.

The treatment of a free sugar with trichloroacetonitrile in dichloromethane in the presence of a suitable base gives rise to anomerically pure, stable trichloroacetimidates:

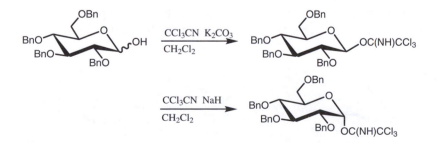

The use of potassium carbonate favours the formation of the β-anomer whereas sodium hydride and DBU favour the α-anomer; prolonged reaction times also result in the formation of the more stable α-anomer.[4] Although Schmidt invokes kinetic/thermodynamic control to explain these results, it may just be that, in the presence of the weaker base (potassium carbonate), a sufficiently high rate of mutarotation of the free sugar and low rate of isomerization of the β-trichloroacetimidate will suffice as an alternative:

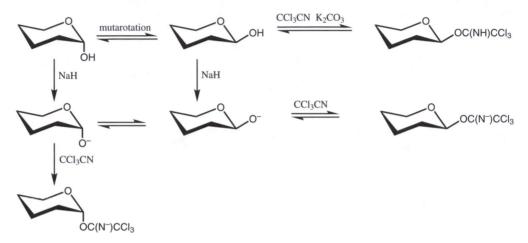

The versatility of the TCA method was obvious from the start:

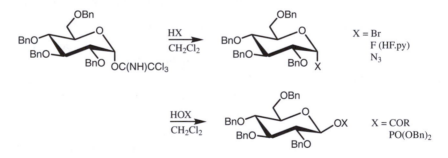

The stronger acids give the thermodynamically more stable products, whether through the conventional oxacarbenium ion intermediates or the isomerization of any initially formed β-D-anomer; the weaker acids give the products of inversion.

The real strength of the TCA method lies in the synthesis of the glycosidic linkage:

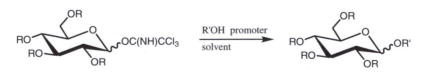

The promoter is generally boron trifluoride etherate or trimethylsilyl triflate (in just catalytic amounts), but zinc bromide,[4] silver(I) triflate[5] and dibutylboron triflate[6] have also been used on occasion:

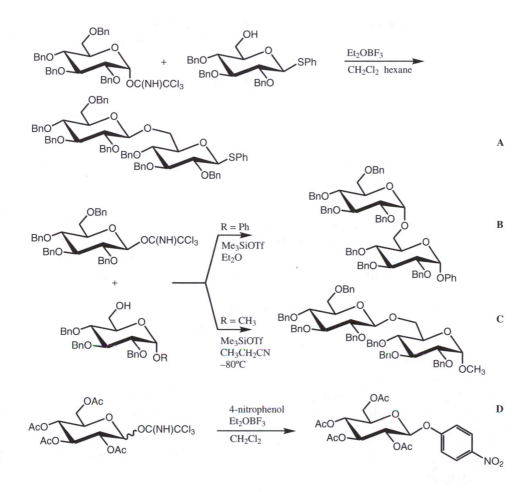

For TCA donors having a non-participating group at C2, treatment with a mild promoter in the presence of an acceptor alcohol generally results in the formation of a glycoside with inversion of configuration at the anomeric carbon (entry **A**), presumably by an S_N2-type process involving an intimate ion-pair. For glycosidations that are promoted by the stronger trimethylsilyl triflate, the fate of the initially formed oxacarbenium ion is regulated by the solvent. In dichloromethane, the stable α-D-glycoside is formed; in ether, the same α-D-glycoside is formed, probably enhanced by the intermediacy of a new β-D-oxonium ion (entry **B**); in acetonitrile or propionitrile, at low temperature, the rapid formation of the sterically unencumbered "α-nitrilium" ion is favoured, leading to the formation of the β-D-glycoside (entry **C**).

For TCA donors having a participating group at C2, the products of glycosidation are generally those expected (entry **D**). A recent paper has indicated that a degree of kinetic control may be exerted in the actual glycosidation step with TCA donors.[7]

Finally, several other features of the TCA method have served to raise its profile among carbohydrate chemists:

- The incidence of competing orthoester formation with TCA donors having a participating group at C2 is low, presumably owing to the acidic (BF_3, TMSOTf or HOTf) nature of the reaction medium (which encourages the rearrangement of unwanted orthoester into the desired glycoside).
- The incidence of acyl group transfer from a TCA donor having a participating group at C2 is low. Although the reason(s) for this remains unclear, it may just be that both boron trifluoride and trimethylsilyl triflate are loath to perform the necessary activation of O2 to initiate the transfer.
- For very unreactive acceptor alcohols, where rearrangement of the TCA donor into an *N*-glycosyl trichloroacetamide can be a problem, Schmidt has developed an "inverse procedure" — the promotor and alcohol are first mixed and then added to the TCA donor.[8,9]
- The method is well suited to the synthesis of glycofuranosides.[10]

References

1. Schmidt, R. R. and Kinzy, W. (1994). *Adv. Carbohydr. Chem. Biochem.*, **50**, 21.
2. Schmidt, R. R. (1996). The anomeric *O*-alkylation and the trichloroacetimidate method – versatile strategies for glycoside bond formation, in *Modern Methods in Carbohydrate Synthesis*, Khan, S. H. and O'Neill, R. A. eds, Harwood Academic, Netherlands, p. 20.
3. Pougny, J.-R., Jacquinet, J.-C., Nassr, M., Duchet, D., Milat, M.-L. and Sinaÿ, P. (1977). *J. Am. Chem. Soc.*, **99**, 6762.
4. Urban, F. J., Moore, B. S. and Breitenbach, R. (1990). *Tetrahedron Lett.*, **31**, 4421.
5. Douglas, S. P., Whitfield, D. M. and Krepinsky, J. J. (1993). *J. Carbohydr. Chem.*, **12**, 131.
6. Wang, Z.-G., Douglas, S. P. and Krepinsky, J. J. (1996). *Tetrahedron Lett.*, **37**, 6985.
7. Kasuya, M. C. and Hatanaka, K. (1998). *Tetrahedron Lett.*, **39**, 9719.
8. Schmidt, R. R. and Toepfer, A. (1991). *Tetrahedron Lett.*, **32**, 3353.

9. Liu, M., Yu, B. and Hui, Y. (1998). *Tetrahedron Lett.*, **39**, 415.
10. Gelin, M., Ferrieres, V. and Plusquellec, D. (1997). *Carbohydr. Lett.*, **2**, 381.

Thioglycosides[1] (1,2-*cis* and 1,2-*trans*)

Thioglycosides, where a sulfur atom replaces the oxygen of the aglycon, are stable derivatives of carbohydrates; we have already addressed the preparation of such molecules (usually of the β-D-configuration), and other methods exist.[2]

We have also had a glimpse of the versatility of thioglycosides in their conversion into glycosyl bromides and glycosyl fluorides. In fact, glycosyl bromides may be generated from thioglycosides *in situ* and used directly for the conventional synthesis of 1,2-*cis* (halide catalysis) and 1,2-*trans* (Koenigs–Knorr) glycosides:[3,4]

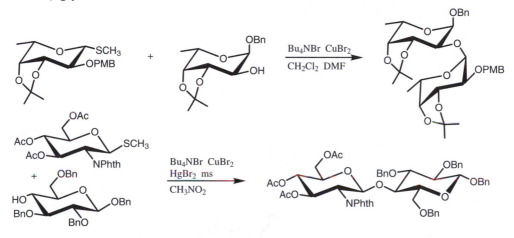

Another reagent, dimethyl(methylthio)sulfonium triflate, can also initiate the "halide catalysis" cascade on a thioglycoside:[5]

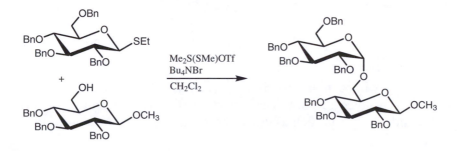

The real strength of thioglycosides in oligosaccharide synthesis is that, because of their weak basicity and low reactivity, they are capable of surviving the effect of most of the promoters that are used in the other glycosylation protocols; thus, they offer a measure of "temporary protection" to one anomeric centre. Indeed, after the initial observation by Ferrier,[6] a whole range of promoters has been developed specifically for the activation of thioglycosides as glycosyl donors:

CH$_3$OTf	NIS, TfOH
Me$_2$S(SMe)OTf	IDCClO$_4$ or IDCOTf
CH$_3$SOTf	PhSePhth or PhIO, Mg(ClO$_4$)$_2$[7]
PhSeOTf	(4-BrC$_6$H$_4$)$_3$N$^{.+}$SbCl$_6^{-}$ [8–10]

Of these promoters, methyl triflate is a potent alkylating agent and, therefore, a carcinogen — its use should be avoided.

The general process of glycosylation is straightforward and follows the principles already established with glycosyl halides and the TCA method:

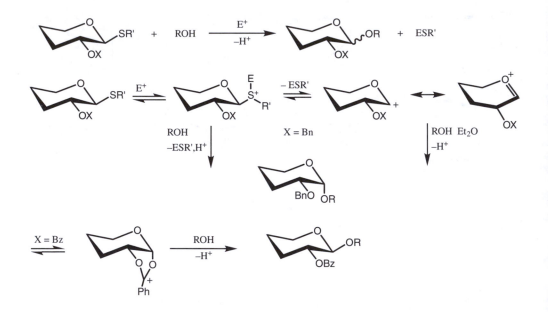

Thioglycosides with a non-participating (benzyl ether) group at C2 generally yield the 1,2-*cis* glycoside, whereas a participating (ester) group gives the 1,2-*trans* glycoside. Solvents such as ether and acetonitrile again tend to favour the formation of 1,2-*cis* and 1,2-*trans* glycosides, respectively. Acetyl group transfer from the donor to the acceptor can again be a problem.

During van Boom's work on thioglycosides, it became apparent that "ether-protected" donors could be activated with iodonium dicollidine perchlorate but "ester-protected" donors remained inert; for the latter type of donor, a "stronger" promotor, *N*-iodosuccinimide/triflic acid was necessary:

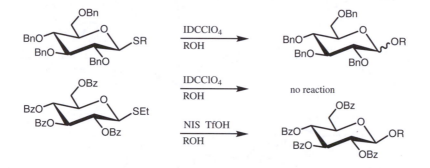

Drawing on Fraser-Reid's earlier concept, van Boom referred to the "ethers" as "armed" thioglycosides and to the "esters" as being correspondingly "disarmed". This concept has proven to be of great value to the synthetic chemist:[11,12]

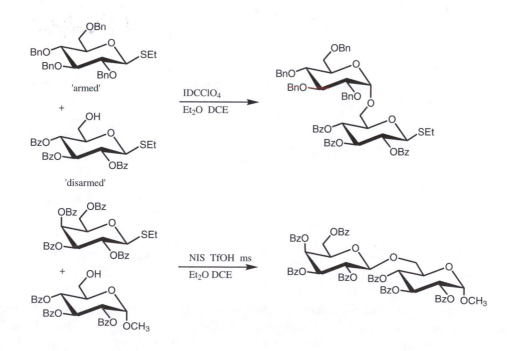

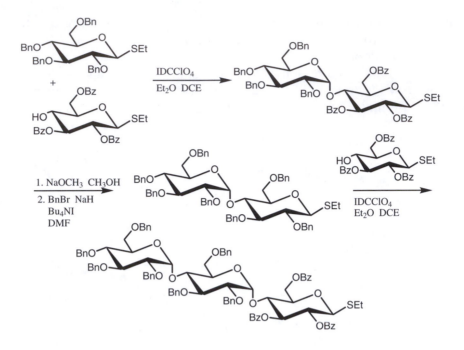

It has been suggested that iodine may be a useful promoter of "armed" thioglycosides[13] and that 2-O-pivaloyl thioglycosides seem particularly suited for the preparation of 1,2-*trans* glycosides.[14,15]

Other factors may be brought into play to adjust the reactivity and selectivity of a thioglycoside, namely the choice of solvent,[16] the size of alkyl groups attached to the sulfur,[17] the anomeric configuration of the thioglycoside,[17] the nature of aromatic groups attached to the sulfur[18,19] and the presence of cyclic protecting groups:

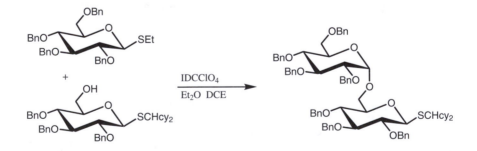

In the matter of cyclic protecting groups, Ley has made something of an art form out of the use of various glycosyl donors, including thioglycosides, and the concepts of "torsional control/reactivity tuning" for glycoside synthesis.[20-23]

Seleno- and Telluro-glycosides[24,25]

Selenoglycosides are prepared from modified carbohydrates in much the same way as are thioglycosides.[26] An anomeric acetate is treated with a Lewis acid and the selenol (a smelly affair!)[27] or a glycosyl bromide is treated with an alkali metal selenide;[28] in general, the choice of starting material often dictates the formation of the β-D-anomer:

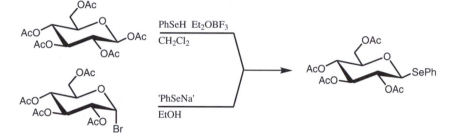

For the corresponding telluroglycosides, it is generally more convenient to proceed *via* the alkali metal telluride, normally generated from the ditelluride:[28]

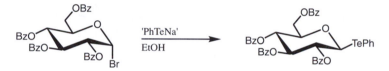

Both seleno- and telluro-glycosides are generally stable, crystalline solids that can be activated into the role of glycosyl donor by treatment with very mild promotors:

AgOTf, K_2CO_3 [27]	IDCClO$_4$ [30]
NIS [29]	NIS, TfOH [30]

This activation is chemoselective — telluroglycosides react in preference to selenoglycosides that, again, are more reactive than thioglycosides.[31] Ley, in concert with the concept of "reactivity tuning", has introduced three different levels of reactivity with the following molecules:[20–23]

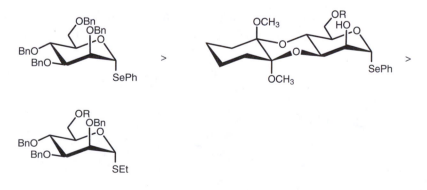

Some examples show the versatility of seleno- and telluro-glycosides in glycoside synthesis:[27,31–33]

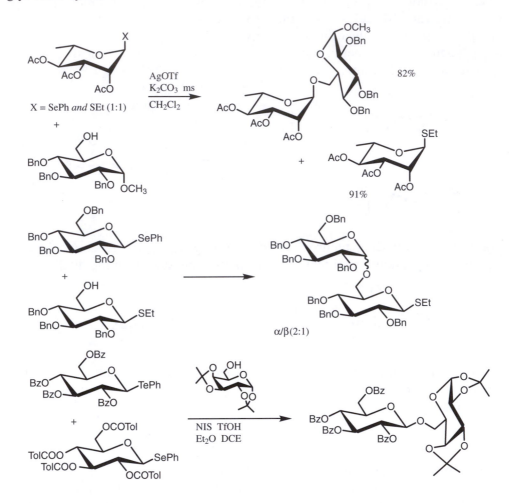

Glycosyl Sulfoxides (sulfinyl glycosides; 1,2-*cis* and 1,2-*trans*)

Thioglycosides, being thioacetals, are naturally amenable to oxidation to produce either the sulfoxide (sulfinyl glycoside) or sulfone (sulfonyl glycoside):

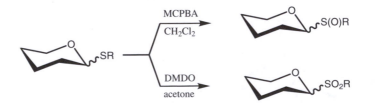

The reagent of choice for the preparation of the sulfone is dimethyldioxirane; the less reactive 3-chloroperbenzoic acid is preferred for the preparation of the sulfoxide.[34,35] In fact, the α-D-thioglycoside may be selectively oxidized to form just one stereoisomer of the sulfoxide (ostensibly owing to a combination of steric and anomeric effects) — the β-D-anomer does not show this partiality:[36]

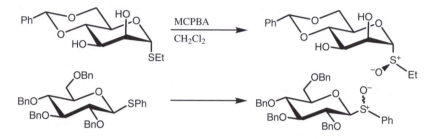

With glycosyl sulfoxides being so available, it was not surprising that they should be investigated as glycosyl donors. However, it took until 1989 for Kahne to realize their potential.[37] The strength of the method is that some very unreactive acceptors may be glycosylated under very mild conditions:[38-41]

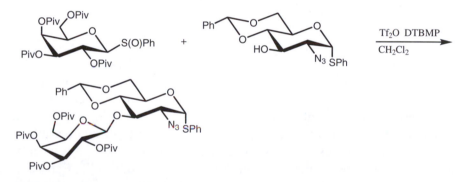

The method seems to rely greatly on the presence of a pivaloyl group at O2 of the donor and this gives rise to the formation of 1,2-*trans* glycosides.[42] However, donors with a benzyl ether at O2 react sufficiently well to provide a synthesis of 1,2-*cis* glycosides.[38]

With triflic anhydride as the promotor, the by-products of the "sulfoxide method" are triflic acid and the very "thiophilic" phenylsulfenyl triflate. A hindered base is very often added to negate the effect of this potentially harmful by-product. As an alternative, Kahne introduced the use of a catalytic amount of triflic acid as the promoter; methyl propiolate was then added as a "scavenger" of the stoichiometric amount of phenylsulfenic acid formed.[40] Other promoter/scavenger combinations have been suggested.[43-45]

All of the general principles of glycoside synthesis can be applied to the "sulfoxide method" and Kahne has been able to "tune" the reactivity of the sulfoxide donor by a careful choice of the substituent on sulfur:[41]

In addition, the actual mechanism of the method has received much attention:[46,47]

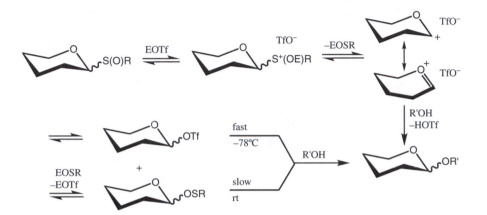

Crich has suggested that glycosyl triflates may well be intermediates in the "sulfoxide method" — at the temperature normally employed ($-78°C$), this appears to be true and could also hold for other methods, for example, a glycosyl bromide and silver(I) triflate. Kahne has shown that glycosyl sulfenates may well be formed early on in the "sulfoxide method" and only by raising the temperature of the reaction mixture later on can a good yield of the glycoside be obtained.

To complete the discussion on sulfoxides as glycosyl donors, it is worthwhile noting two recent procedures ("dehydrative glycosylation") that manage to convert free sugars directly into glycosides, both probably through the intermediacy of a glycosyl triflate:[48,49]

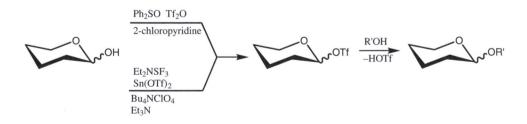

Anomeric "sulfimides" have also been suggested as glycosyl donors.[50]

References

1. Garegg, P. J. (1997). *Adv. Carbohydr. Chem. Biochem.*, **52**, 179.
2. Norberg, T. (1996). Glycosylation properties and reactivity of thioglycosides, sulfoxides, and other *S*-glycosides: current scope and future prospects, in *Modern Methods in Carbohydrate Synthesis*, Khan, S. H. and O'Neill, R. A. eds, Harwood Academic, Netherlands, p. 82.
3. Ludewig, M. and Thiem, J. (1998). *Synthesis*, 56.
4. Sato, S., Mori, M., Ito, Y. and Ogawa, T. (1986). *Carbohydr. Res.*, **155**, C6.
5. Andersson, F., Fügedi, P., Garegg, P. J. and Nashed, M. (1986). *Tetrahedron Lett.*, **27**, 3919.
6. Ferrier, R. J., Hay, R. W. and Vethaviyasar, N. (1973). *Carbohydr. Res.*, **27**, 55.
7. Fukase, K., Nakai, Y., Kanoh, T. and Kusumoto, S. (1998). *Synlett*, 84.
8. Sinaÿ, P. (1991). *Pure Appl. Chem.*, **63**, 519.
9. Zhang, Y.-M., Mallet, J.-M. and Sinaÿ, P. (1992). *Carbohydr. Res.*, **236**, 73.
10. Mehta, S. and Pinto, B. M. (1998). *Carbohydr. Res.*, **310**, 43.
11. Veeneman, G. H. and van Boom, J. H. (1990). *Tetrahedron Lett.*, **31**, 275.
12. Veeneman, G. H., van Leeuwen, S. H. and van Boom, J. H. (1990). *Tetrahedron Lett.*, **31**, 1331.
13. Kartha, K. P. R., Aloui, M. and Field, R. A. (1996). *Tetrahedron Lett.*, **37**, 5175.
14. Sato, S., Nunomura, S., Nakano, T., Ito, Y. and Ogawa, T. (1988). *Tetrahedron Lett.*, **29**, 4097.
15. Knapp, S. and Nandan, S. R. (1994). *J. Org. Chem.*, **59**, 281.
16. Demchenko, A., Stauch, T. and Boons, G.-J. (1997). *Synlett*, 818.
17. Geurtsen, R., Holmes, D. S. and Boons, G.-J. (1997). *J. Org. Chem.*, **62**, 8145.
18. Roy, R., Andersson, F. O. and Letellier, M. (1992). *Tetrahedron Lett.*, **33**, 6053.
19. Sliedregt, L. A. J. M., Zegelaar-Jaarsveld, K., van der Marel G. A. and van Boom, J. H. (1993). *Synlett*, 335.
20. Grice, P., Ley, S. V., Pietruszka, J., Priepke, H. W. M. and Walther, E. P. E. (1995). *Synlett*, 781.
21. Grice, P., Ley, S. V., Pietruszka, J., Osborn, H. M. I., Priepke, H. W. M. and Warriner, S. L. (1997). *Chem. Eur. J.*, **3**, 431.
22. Douglas, N. L., Ley, S. V., Lücking, U. and Warriner, S. L. (1998). *J. Chem. Soc., Perkin Trans. 1*, 51.
23. Green, L., Hinzen, B., Ince, S. J., Langer, P., Ley, S. V. and Warriner, S. L. (1998). *Synlett*, 440.
24. Mehta, S. and Pinto, B. M. (1996). Phenyl selenoglycosides as versatile glycosylating agents in oligosaccharide synthesis and the chemical synthesis of disaccharides containing sulfur and selenium, in *Modern Methods in Carbohydrate Synthesis*, Khan, S. H. and O'Neill, R. A. eds, Harwood Academic, Netherlands, p. 107.
25. Witczak, Z. J. and Czernecki, S. (1998). *Adv. Carbohydr. Chem. Biochem.*, **53**, 143.
26. Ferrier, R. J. and Furneaux, R. H. (1980). *Methods Carbohydr. Chem.*, **8**, 251.
27. Mehta, S. and Pinto, B. M. (1993). *J. Org. Chem.*, **58**, 3269.
28. Stick, R. V., Tilbrook, D. M. G. and Williams, S. J. (1997). *Aust. J. Chem.*, **50**, 233.
29. Heskamp, B. M., Veeneman, G. H., van der Marel, G. A., van Boeckel, C. A. A. and van Boom, J. H. (1995). *Tetrahedron*, **51**, 8397.
30. Zuurmond, H. M., van der Meer, P. H., van der Klein, P. A. M., van der Marel, G. A. and van Boom, J. H. (1993). *J. Carbohydr. Chem.*, **12**, 1091.
31. Stick, R. V., Tilbrook, D. M. G. and Williams, S. J. (1997). *Aust. J. Chem.*, **50**, 237.

32. Johnston, B. D. and Pinto, B. M. (1998). *Carbohydr. Res.*, **305**, 289.
33. Yamago, S., Kokubo, K., Murakami, H., Mino, Y., Hara, O. and Yoshida, J.-i. (1998). *Tetrahedron Lett.*, **39**, 7905.
34. Skelton, B. W., Stick, R. V., Tilbrook, D. M. G., White, A. H. and Williams, S. J. (2000). *Aust. J. Chem.*, **53**, 389.
35. Kakarla, R., Dulina, R. G., Hatzenbuhler, N. T., Hui, Y. W. and Sofia, M. J. (1996). *J. Org. Chem.*, **61**, 8347.
36. Crich, D., Mataka, J., Sun, S., Lam, K.-C., Rheingold, A. L. and Wink, D. J. (1998). *Chem. Commun.*, 2763.
37. Kahne, D., Walker, S., Cheng, Y. and Van Engen, D. (1989). *J. Am. Chem. Soc.*, **111**, 6881.
38. Yan, L. and Kahne, D. (1996). *J. Am. Chem. Soc.*, **118**, 9239.
39. Ikemoto, N. and Schreiber, S. L. (1990). *J. Am. Chem. Soc.*, **112**, 9657.
40. Yang, D., Kim, S.-H. and Kahne, D. (1991). *J. Am. Chem. Soc.*, **113**, 4715.
41. Raghavan, S. and Kahne, D. (1993). *J. Am. Chem. Soc.*, **115**, 1580.
42. Thompson, C., Ge, M. and Kahne, D. (1999). *J. Am. Chem. Soc.*, **121**, 1237.
43. Sliedregt, L. A. J. M., van der Marel, G. A. and van Boom, J. H. (1994). *Tetrahedron Lett.*, **35**, 4015.
44. Alonso, I., Khiar, N. and Martín-Lomas, M. (1996). *Tetrahedron Lett.*, **37**, 1477.
45. Gildersleeve, J., Smith, A., Sakurai, K., Raghaven, S. and Kahne, D. (1999). *J. Am. Chem. Soc.*, **121**, 6176.
46. Crich, D. and Sun, S. (1997). *J. Am. Chem. Soc.*, **119**, 11217.
47. Gildersleeve, J., Pascal, R. A., Jr. and Kahne, D. (1998). *J. Am. Chem. Soc.*, **120**, 5961.
48. Garcia, B. A. and Gin, D. Y. (2000). *J. Am. Chem. Soc.*, **122**, 4269.
49. Hirooka, M. and Koto, S. (1998). *Bull. Chem. Soc. Jpn.*, **71**, 2893.
50. Cassel, S., Plessis, I., Wessel, H. P. and Rollin, P. (1998). *Tetrahedron Lett.*, **39**, 8097.

Glycals and 4-Pentenyl Glycosides (1,2-*cis* and 1,2-*trans*)

Glycals and 4-pentenyl glycosides will be discussed successively here because, for somewhat different reasons, these two alkenes are capable of acting as glycosyl donors:

Glycals

The treatment of a suitable glycal with a Lewis acid in the presence of a reactive alcohol gives rise to glycosides having unsaturation in the ring:[1,2]

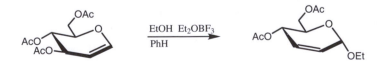

The Lewis acid coordinates to the good leaving group at O3 and a resonance-stabilized cation results:

This cation is attacked by the alcohol to yield the 2,3-unsaturated glycoside.

Although the chemistry of glycals was to flourish in the twenty years after Ferrier's discovery, as discussed above, it took two significant events to establish these unsaturated sugars as glycosyl donors in their own right: first, the availability of dimethyldioxirane as a laboratory reagent and, second, the attraction of Samuel Danishefsky to the field of carbohydrates.

A glycal does not seem to be the ideal glycosyl donor. Apart from the problem of stereoselectivity at the anomeric carbon (α- or β-), an oxygen atom must be reinstated at C2, again stereoselectively; the generation of an intermediate epoxide solved both of these problems:[3, 4]

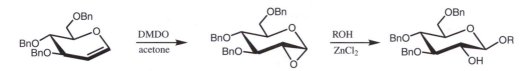

One of the many advantages of this method was that the initial glycoside formed possessed a free hydroxyl group at C2, available for further elaboration:

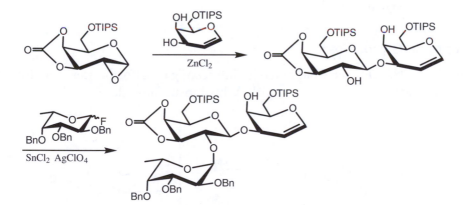

Several points need to be made about this "glycal epoxide" methodology:

- Only DMDO can be used to generate the epoxide from the glycal — other reagents, such as MCPBA, cause subsequent decomposition of the epoxide.

- The epoxide is usually formed with high stereoselectivity, being installed in a "*trans*" sense to the substituent at C3.
- Normally, an "α-D-" epoxide will yield a β-D-glycoside upon treatment with an alcohol and a Lewis acid. Sometimes, α-D-glycosides result with the use of less reactive alcohols.

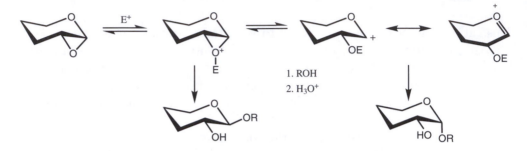

- When an anomeric epoxide is unsatisfactory as a glycosyl donor, conventional transformations may yield a more suitable donor.

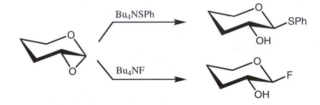

We will return to this methodology later in the book when we consider the concept of "glycal assembly" for the synthesis of a breast tumour antigen.

4-Pentenyl Glycosides[5-8]

In the late 1980s, Fraser-Reid noted an interesting transformation of a "higher" sugar derivative upon treatment with *N*-bromosuccinimide:

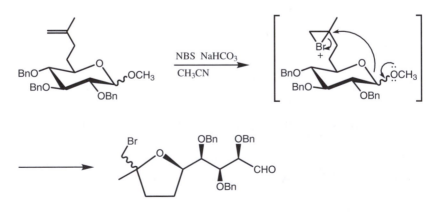

This chance observation led to the idea that 4-pentenyl glycosides would respond similarly to NBS, thus providing a new type of protecting group for the anomeric centre:

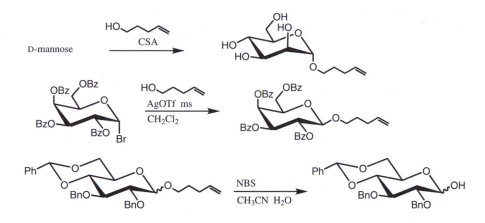

In addition, the 4-pentenyl glycosides were found to be effective glycosyl donors:

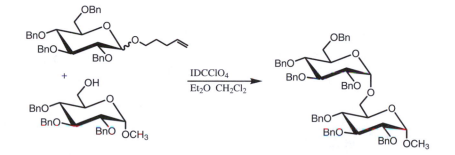

So was born the "NPG" (*n*-pentenyl glycoside) method of glycosylation, a method that has gained in popularity because of the ease of installation of the 4-pentenyl glycoside, the stability of the group to most reagents (much akin to a thioglycoside) and the easy promotion by oxidizing agents such as NBS and $IDCClO_4$.

The NPG method was to be a fertile area for Fraser-Reid during the next decade — out of it was to come the concept of "armed/disarmed" glycosyl donors and the necessity for corresponding promoters:

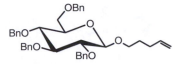

armed – $IDCClO_4$ promotion

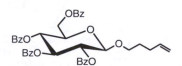

disarmed – NIS, TfOH or NIS, Et_3SiOTf promotion

A whole new facet of glycoside synthesis was exposed:

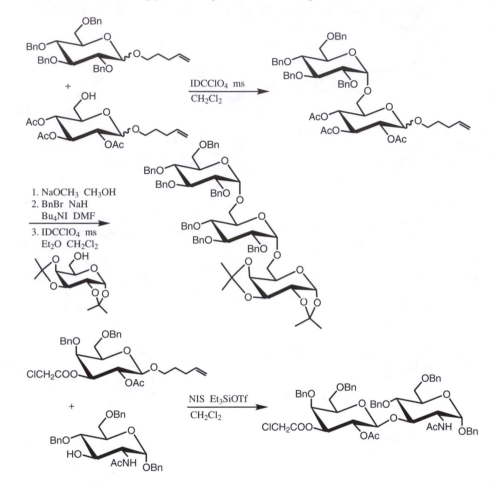

A corollary of the NPG method is that an armed donor may be "side-tracked" (protected) by conversion into a vicinal dibromide; when necessary, the dibromide (a "latent" glycosyl donor) may be treated with zinc metal to regenerate the armed donor:[9]

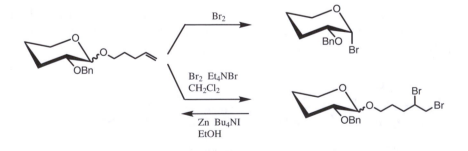

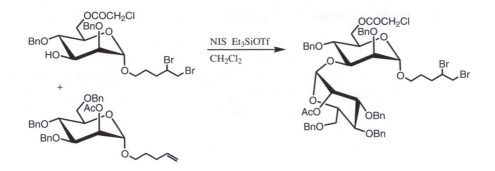

The mechanism of the NPG method has been investigated in some depth and is fairly well understood:

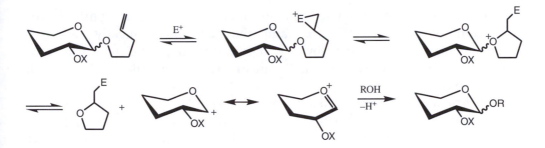

The stereochemical outcome depends, as usual, on the nature of "X" — ethers generally give 1,2-*cis* glycosides and esters the 1,2-*trans* glycoside.

References

1. Ferrier, R. J. and Prasad, N. (1969). *J. Chem. Soc. C*, 570.
2. Ferrier, R. J. (1969). *Adv. Carbohydr. Chem. Biochem.*, **24**, 199.
3. Danishefsky, S. J. and Bilodeau, M. T. (1996). *Angew. Chem. Int. Ed. Engl.*, **35**, 1380.
4. Seeberger, P. H., Bilodeau, M. T. and Danishefsky, S. J. (1997). *Aldrichimica Acta*, **30**, 75.
5. Fraser-Reid, B., Udodong, U. E., Wu, Z., Ottosson, H., Merritt, J. R., Rao, C. S., Roberts, C. and Madsen, R. (1992). *Synlett*, 927.
6. Fraser-Reid, B., Merritt, J. R., Handlon, A. L. and Andrews, C. W. (1993). *Pure Appl. Chem.*, **65**, 779.
7. Fraser-Reid, B. and Madsen, R. (1997). Oligosaccharide synthesis by *n*-pentenyl glycosides, in *Preparative Carbohydrate Chemistry*, Hanessian, S. ed., Marcel Dekker, New York, p. 339.
8. Madsen, R. and Fraser-Reid, B. (1996). *n*-Pentenyl glycosides in oligosaccharide synthesis, in *Modern Methods in Carbohydrate Synthesis*, Kahn, S. H. and O'Neill, R. A. eds, Harwood Academic, Netherlands, p. 155.
9. Rodebaugh, R., Debenham, J. S., Fraser-Reid, B. and Snyder, J. P. (1999). *J. Org. Chem.*, **64**, 1758.

β-D-Mannopyranosides[1,2] (1,2-*cis*)

D-Mannose is a common constituent of many naturally occurring oligosaccharides, some of which are attached to proteins and all of which play a biological role in the host organism. Although the common linkage in such oligosaccharides is α-D-*manno*, there is a frequent enough occurrence of the β-D-counterpart. For example, the pentasaccharide core of *N*-linked glycoproteins is invariant and contains both types of linkages:

Although we have discussed many methods for the synthesis of α-D-mannopyranosides (1,2-*trans* and the thermodynamically favoured arrangement), only a very few of these are adaptable to form β-D-mannopyranosides. Here, we shall indicate these versatile methods and, also, introduce new ones.

The origin of the problem in any approach to β-D-mannopyranosides is the axial orientation of the group at C2. As we have seen, if this group is an ester, then the α-D-mannopyranoside generally results from participation of the ester at the anomeric carbon; if the group is an ether, then a favourable anomeric effect again results in the formation of the α-D-anomer:

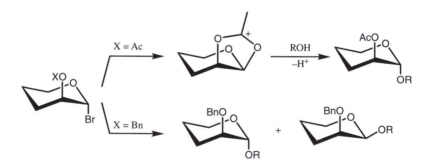

Obviously, to circumvent these problems, some creative thinking was required!

Glycosyl Halides

Probably one of the most common methods for the synthesis of β-D-mannopyranosides involves the treatment of a 2-*O*-benzyl-α-D-mannopyranosyl halide with an insoluble promoter such as silver(I) silicate or silver(I) zeolite:

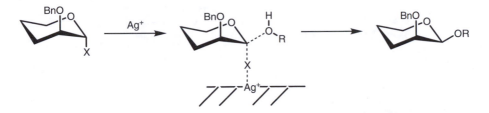

The strength of this Koenigs–Knorr approach lies in its simplicity and it is successful because the "push-pull" type of mechanism invoked on the surface of the promoter effectively shields the α-face of the donor, forcing the acceptor alcohol to approach in the required manner:[3-5]

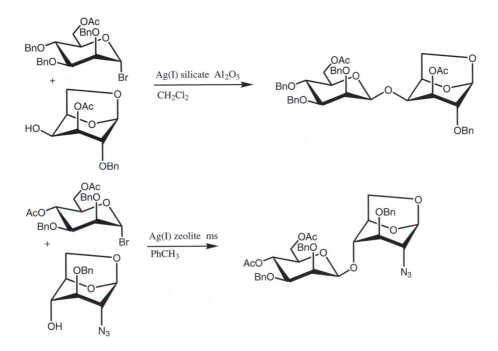

Glycosyl Sulfoxides

The seminal paper by Crich suggested that glycosyl triflates were intermediates in the "sulfoxide method" of glycosidation.[6] Indeed, as applied to α-D-mannopyranosyl sulfoxides having non-participating groups at C2 and C3 and a 4,6-*O*-benzylidene protecting group (presumably to discourage glycosyl cation formation), there now exists a powerful method for the synthesis of β-D-mannopyranosides from virtually any sort of acceptor alcohol:[7–9]

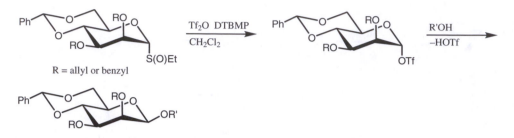

R = allyl or benzyl

The stereoselectivity of the method is excellent and, in what may become a general improvement, even 1-thio-α-D-mannopyranosides were activated by phenylsulfenyl triflate to provide β-D-mannopyranosides:[10]

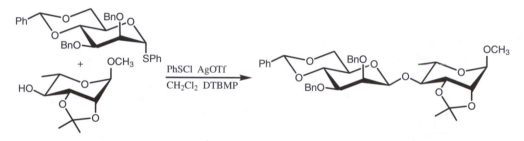

β-D-Glucopyranoside to β-D-Mannopyranoside

In principle, it should be possible to generate a β-D-glucopyranoside by conventional means and, if O2 is differentially protected, invert the stereochemistry by an oxidation–reduction sequence:

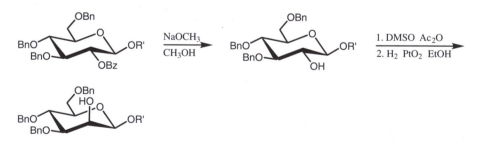

This method of synthesis of β-D-mannopyranosides was first announced in 1972 and has been used on many occasions but it obviously suffers from a major drawback — the preparation of the 2-O-acyl glycosyl donor.[11] A related approach, which inverts the configuration at C2 of the β-D-glucopyranoside by a nucleophilic displacement, again suffers from the early manipulations in the synthesis of the glycosyl donor:[12]

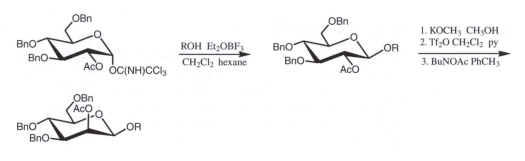

An intramolecular version of this process, which somewhat curiously still requires the presence of a 4,6-*O*-benzylidene group, has been developed by Kunz:[13]

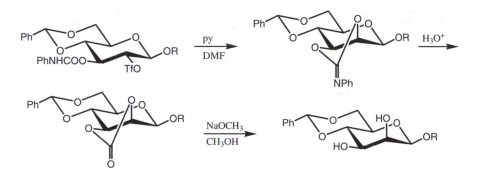

Lichtenthaler has developed a procedure that utilizes glycosyl halides in tandem with a stereoselective reduction for the synthesis of β-D-mannopyrano-sides:[14]

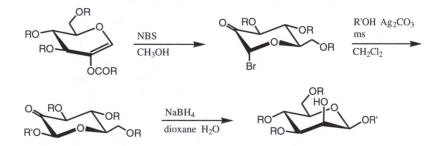

Central to this procedure are the easy preparation of the 2-ulosyl α-D-glycosyl bromide, the β-specific glycosidation (presumably successful owing to suppression of glycosyl cation formation by the adjacent carbonyl group) and the stereoselective reduction.

Intramolecular Aglycon Delivery[15]

This last method to be discussed is certainly the most elegant; whether the extra early steps are compensated for by the final, completely stereoselective glycosidation is a moot point. First announced by Hindsgaul[16] and later modified by Stork[17] and Ogawa and Ito,[18] the method requires the generation of a D-mannopyranosyl donor with the acceptor *already attached* as some sort of acetal at O2; the addition of the appropriate promoter then gives the β-D-mannopyranoside:

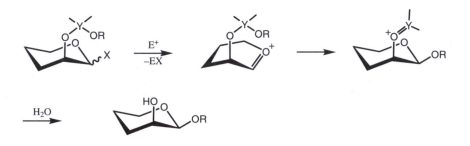

Details of each modification are presented below:[19]

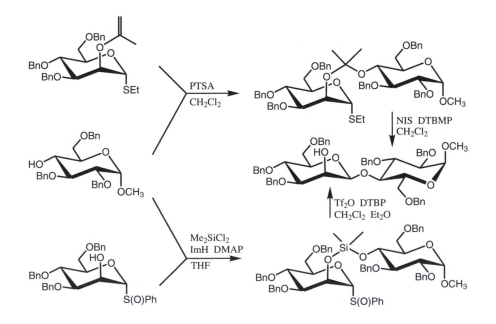

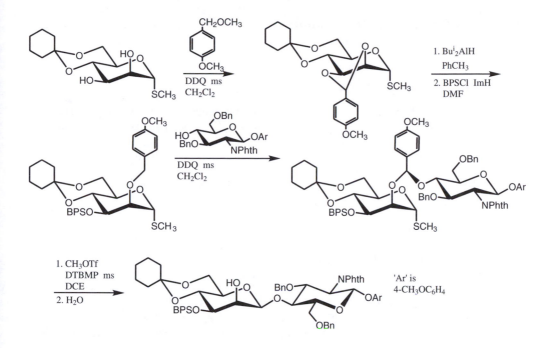

Other Methods

Although other methods have been disclosed for the synthesis of β-D-mannopyranosides, none seem to compete consistently with those discussed above.[1] Of late, methods involving the alkylation of *cis*-1,2-stannylene acetals derived from D-mannose and L-rhamnose,[20] the selective reduction of anomeric orthoesters,[21] the use of a "solid acid" (sulfated zirconia) on a D-mannosyl fluoride,[22] the involvement of "prearranged glycosides"[23] and the double inversion of a β-D-galactopyranoside ditriflate[24] show promise for the synthesis of β-D-manno- and β-L-rhamnopyranosides. One of the rare instances of the use of the Mitsunobu reaction for the synthesis of glycosides was announced by Garegg. The method is best for the preparation of aryl β-D-mannopyranosides, mainly owing to the anomeric purity of the starting hemiacetal, the general "inversion of configuration" associated with the process and the appropriate acidity of the glycosyl acceptor:[25, 26]

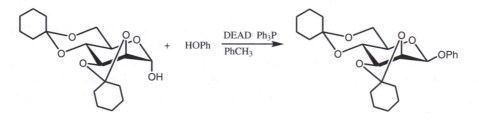

The method is general for a whole range of substituents on the aromatic ring.[27]

References

1. Barresi, F. and Hindsgaul, O. (1996). Synthesis of β-D-mannose containing oligosaccharides, in *Modern Methods in Carbohydrate Synthesis*, Khan, S. H. and O'Neill, R. A. eds, Harwood Academic, Netherlands, p. 251.
2. Gridley, J. J. and Osborn, H. M. I. (2000). *J. Chem. Soc., Perkin Trans. 1*, 1471.
3. Paulsen, H. and Lockhoff, O. (1981). *Chem. Ber.*, **114**, 3102.
4. van Boeckel, C. A. A., Beetz, T. and van Aelst, S. F. (1984). *Tetrahedron*, **40**, 4097.
5. Garegg, P. J. and Ossowski, P. (1983). *Acta Chem. Scand.*, **B37**, 249.
6. Crich, D. and Sun, S. (1997). *J. Am. Chem. Soc.*, **119**, 11217.
7. Crich, D. and Sun, S. (1998). *Tetrahedron*, **54**, 8321.
8. Crich, D. and Barba, G. R. (1998). *Tetrahedron Lett.*, **39**, 9339.
9. Crich, D. and Dai, Z. (1999). *Tetrahedron*, **55**, 1569.
10. Crich, D., Cai, W. and Dai, Z. (2000). *J. Org. Chem.*, **65**, 1291.
11. Ekborg, G., Lindberg, B. and Lönngren, J. (1972). *Acta Chem. Scand.*, **26**, 3287.
12. Fürstner, A. and Konetzki, I. (1998). *Tetrahedron Lett.*, **39**, 5721.
13. Günther, W. and Kunz, H. (1992). *Carbohydr. Res.*, **228**, 217.
14. Lichtenthaler, F. W., Kläres, U., Szurmai, Z. and Werner, B. (1998). *Carbohydr. Res.*, **305**, 293.
15. Garegg, P. J. (1992). *Chemtracts-Org. Chem.*, **5**, 389.
16. Barresi, F. and Hindsgaul, O. (1994). *Can. J. Chem.*, **72**, 1447.
17. Stork, G. and La Clair, J. L. (1996). *J. Am. Chem. Soc.*, **118**, 247.
18. Ito, Y., Ohnishi, Y., Ogawa, T. and Nakahara, Y. (1998). *Synlett*, 1102.
19. Ennis, S. C., Fairbanks, A. J., Tennant-Eyles, R. J. and Yeates, H. S. (1999). *Synlett*, 1387.
20. Hodosi, G. and Kováč, P. (1998). *Carbohydr. Res.*, **308**, 63.
21. Ohtake, H., Iimori, T. and Ikegami, S. (1998). *Synlett*, 1420.
22. Toshima, K., Kasumi, K.-i. and Matsumura, S. (1998). *Synlett*, 643.
23. Ziegler, T. and Lemanski, G. (1998). *Angew. Chem. Int. Ed.*, **37**, 3129.
24. Sato, K.-i. and Yoshitomo, A. (1995). *Chem. Lett.*, 39.
25. Åkerfeldt, K., Garegg, P. J. and Iversen, T. (1979). *Acta Chem. Scand.*, **B33**, 467.
26. Roush, W. R. and Lin, X.-F. (1991). *J. Org. Chem.*, **56**, 5740.
27. Zechel, D. and Withers, S. G., unpublished results.

Miscellaneous Methods, and *C*-Glycosides

Miscellaneous Methods

The glycosidic linkage has exerted a certain power over chemists for centuries — it has never yielded to just one method of synthesis and it has offered a sort of "fatal attraction" for improvements in its construction. So, in addition to the tried-and-tested procedures discussed so far, there exists an array of other methods that are either in their infancy or, as yet, have not gained general acceptance.

Glycosyl esters: Certainly, glycosyl acetates are easily prepared and, when treated with a Lewis acid and a phenol, give rise to aryl glycosides.[1] Mukaiyama has done much to extend the method for a useful synthesis of 1,2-*cis* glycosides:[2]

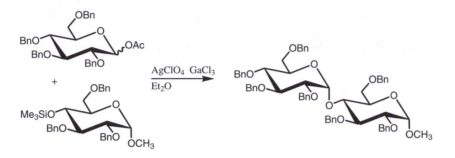

Iron(III) chloride has recently been used in a synthesis of 1,2-*cis* glycosides *having a participating group at C2*. The authors suggest that such a product arises from an anomerization of the originally formed 1,2-*trans* glycoside:[3]

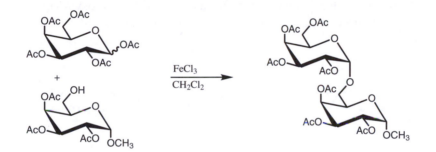

Glycosyl xanthates (*O*-alkyl *S*-glycosyl dithiocarbonates) have been touted as the glycosyl donor of choice for the synthesis of α-D-sialosides:[4,5]

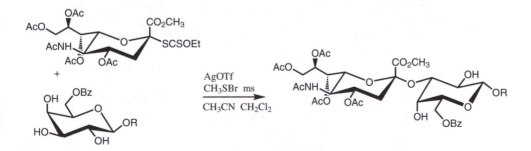

Sinaÿ has significantly extended the method.[6,7]

Various esters derived from some of the phosphorus oxyacids have been developed as useful glycosyl donors. So far, *O*-glycosyl phosphates[8–10] and phosphites have proved the most popular, with the latter being preferred owing to the easier preparation and ready activation:[11-14]

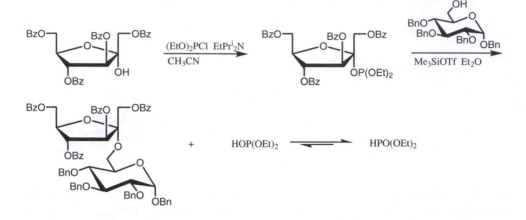

Alkenyl glycosides: We have already encountered alkenyl glycosides in our discussion of the "latent/active" concept as applied to glycosyl donors. In fact, there is a range of such donors that have been used successfully in the synthesis of glycosides:[6,7,15-18]

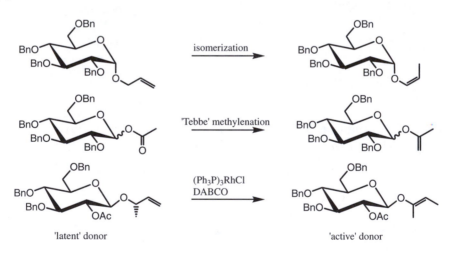

Two interesting variations on the same theme are a glycosyl isopropenyl carbonate and an isopropenyl ether on the acceptor:[6,7,15]

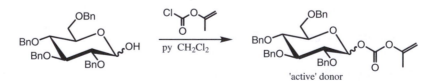

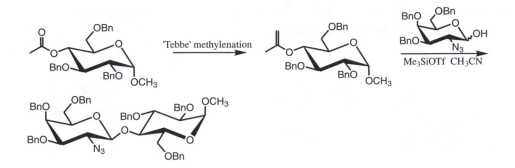

All of the above processes operate on much the same principle:

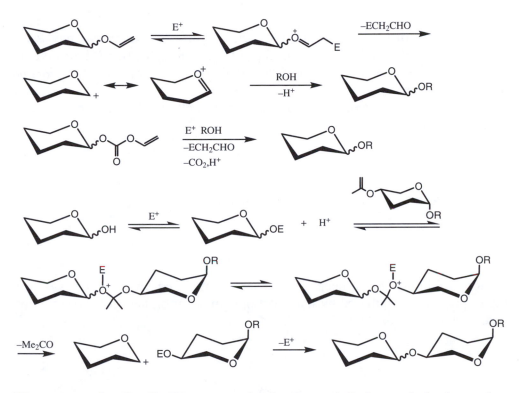

"Remote activation": "Remote activation", as applied to carbohydrates, has essentially been the brainchild of Hanessian.[19] In essence, many methods of forming the glycosidic bond involve direct activation of the atom attached to the anomeric carbon:

'X' is OH, halogen, SR

Other methods, however, rely on the activation of a group that is remote from the anomeric carbon but attached to it *via* other atoms:

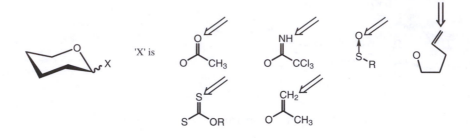

Although this distinction, with these two groups of examples, may seem somewhat forced and flimsy, the concept of "remote activation" does take on some significance when one attempts to develop a glycosyl donor *that bears no protecting groups*.

The first generation of such molecules requiring "remote activation" were 2-pyridylthio glycosides:[20]

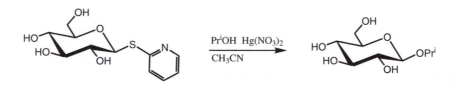

Although mixtures of anomers were generally obtained, the reaction was very rapid, presumably initiated by some sort of complexation between the metal ion and the bidentate aglycon:

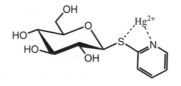

Later, the method was extended to include such donors as 3-methoxy-2-pyridyloxy (MOP) glycosides that, although activated by just a "catalytic"

amount of methyl triflate (and thereafter the liberated triflic acid), still required the use of a large excess of the glycosyl acceptor:

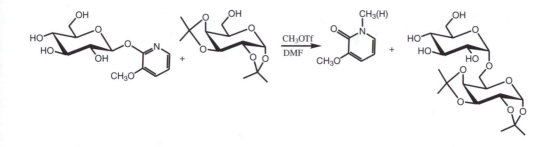

Generally, the S_N2 nature of the process guaranteed the preponderance of the 1,2-*cis* glycoside. Further studies by Hanessian showed that *protected* MOP donors, as with most other sorts of glycosyl donors, could be used effectively for the glycosylation of *equimolar* amounts of acceptor alcohols — the ideal promoter was copper(II) triflate. The stereochemical outcome of the various glycosidations depended, predictably, on the nature of the protecting group at O2 and on the solvent.

A further development in the "remote activation" concept was the use of *O*-glycosyl *S*-(2-pyridyl) thiocarbonates (TOPCAT) as glycosyl donors:[19]

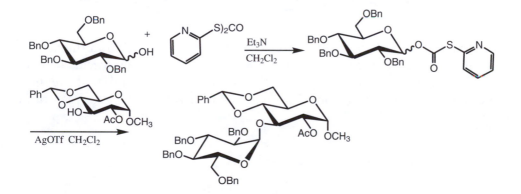

The unique promoter, silver(I) triflate, again probably operating through the formation of some sort of bidentate species, allows for the chemoselective activation of TOPCAT over MOP [copper(II) triflate] groups.

More recently, Hanessian has applied the "MOP glycoside" approach to the synthesis of glycosyl phosphates and their nucleoside derivatives[21] and Kobayashi has expanded upon the use of "glycosyl 2-pyridinecarboxylates" for the preparation of various disaccharides:[22]

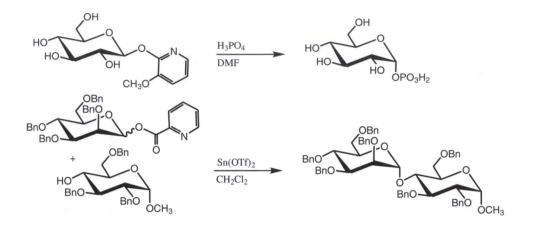

C-Glycosides[23–31]

We have spent a deal of time discussing glycosides and, *inter alia*, N- and S-glycosides:

These three classes of compounds are actually acetals, aminals and thioacetals, all relatively stable to the action of bases but generally unstable to acid, acid and thiophilic reagents, respectively. Another class of compound that we have neglected until now arises from the replacement of the exocyclic oxygen of an acetal by carbon:

These compounds are known as C-glycosides; in fact, they are derivatives of tetrahydropyran and, as such, are simply cyclic ethers. Consequently, they are inert compounds, essentially stable to the action of both acids and bases.

Why, then, are C-glycosides of interest? First, many natural products contain a C-glycosidic linkage: aquayamycin is a C-aryl glycoside produced by *Streptomyces misawanensis* and possesses antibiotic activity,[32] showdomycin is a C-nucleoside possessing antibacterial and antitumour properties[23] and palytoxin and maitotoxin[33] are two marine metabolites that are among some of the most toxic chemicals known.

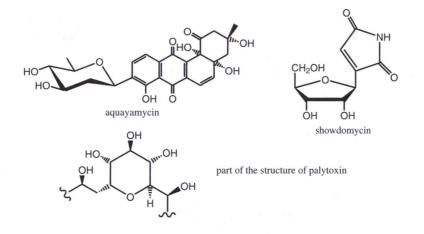

aquayamycin

showdomycin

part of the structure of palytoxin

Second, chemists have long held the notion that C-glycosides would be stable "mimics" of glycosides and so be potential candidates for the design of molecules with expected biological activity, but possessing none of the lability of a glycoside.[28, 34] Whatever the reason, there has been an enormous drive over the past two decades for the synthesis of C-glycosides.

Whereas the synthesis of glycosides (C–O bond construction), by and large, is restricted to methods that develop a partial positive charge at the anomeric carbon atom, such is not the case with the synthesis of C-glycosides — the construction of a C–C bond allows a deal more flexibility in the method. The three main routes followed involve both ionic and free radical processes:

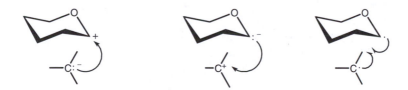

Again, stereoselectivity of bond formation at the anomeric carbon (α or β) is of prime importance.

The addition of carbanions to anomeric electrophiles: There is a plethora of different methods for the addition of a carbanion to an anomeric electrophile and only a few will be mentioned here. The addition of an organometallic reagent to a glycosyl halide is one of the older methods but yields, especially with ester protecting groups in the sugar, are often low:

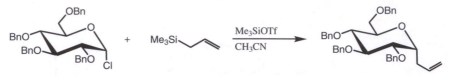

Of general usage is the method developed by Kishi — the addition of an organometallic reagent to an aldonolactone:[35,36]

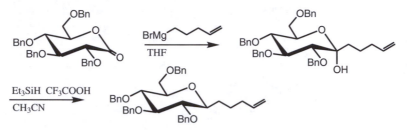

Epoxides are useful alkylating agents of organometallic species:[35,37]

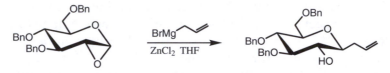

Free sugars, upon chain extension with phosphoranes, phosphonates or organometallic species, yield alkenes which cyclize either spontaneously or by treatment with bases or electrophiles:

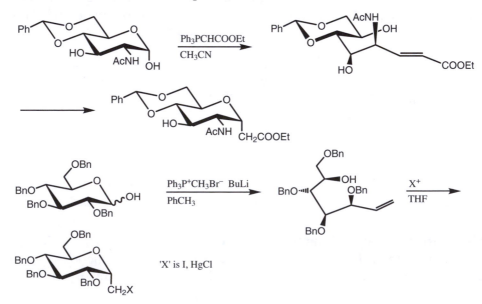

The addition of electrophiles to anomeric carbanions: Again, many methods exist for the addition of an electrophile to an anomeric carbanion but most typically utilize the presence of an electron-withdrawing group at the anomeric carbon to aid in the generation and stabilization of the negative charge:[38]

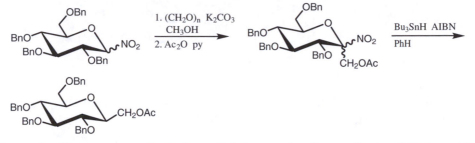

Glycosyl sulfones are particularly useful donors for the synthesis of *C*-glycosides, offering versatility not found with other groups:[39–41]

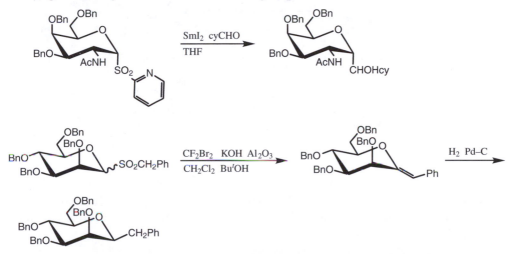

Very often, when the anion is not stabilized by an electron-withdrawing group at the anomeric carbon, the presence of an oxygen atom at C2 will cause a β-elimination; this can either be avoided or put to good use:[42–44]

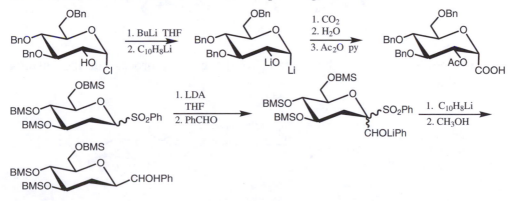

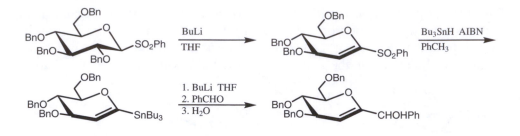

Glycosyl radicals:[45-48]

In the early 1980s, it was first noticed that anomeric radicals add to electron-deficient alkenes to give axial *C*-glycosides:

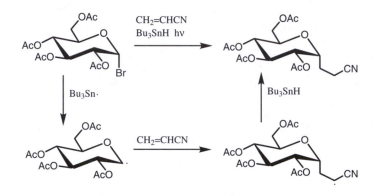

Although the process can sometimes be plagued by side reactions of the highly reactive, intermediate glycosyl radical (reduction, hydrogen atom abstraction), it is this very intermediate that holds the key to the high stereoselectivity usually observed:

more reactive ($B_{2,5}$)

Many modifications have been made to the general process in the ensuing twenty years:[49-52]

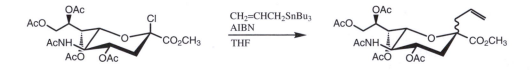

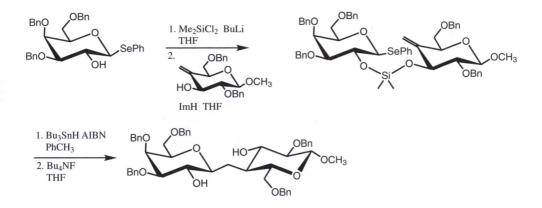

Other methods: Various other methods exist for the synthesis of *C*-glycosides and include substrates such as glycosyl diazirines,[53] glycals,[54] glycosenes[55] and telluroglycosides.[56]

References

1. Conchie, J., Levvy, G. A. and Marsh, C. A. (1957). *Adv. Carbohydr. Chem.*, **12**, 157.
2. Mukaiyama, T., Katsurada, M. and Takashima, T. (1991). *Chem. Lett.*, 985.
3. Chatterjee, S. K. and Nuhn, P. (1998). *Chem. Commun.*, 1729.
4. Lönn, H. and Stenvall, K. (1992). *Tetrahedron Lett.*, **33**, 115.
5. Marra, A. and Sinaÿ, P. (1990). *Carbohydr. Res.*, **195**, 303.
6. Sinaÿ, P. and Mallet, J.-M. (1996). Synthesis and use of *S*-xanthates, carbohydrate enol ethers and related derivatives in the field of glycosylation, in *Modern Methods in Carbohydrate Synthesis*, Khan, S. H. and O'Neill, R. A. eds, Harwood Academic, Netherlands, p. 130.
7. Sinaÿ, P. (1991). *Pure Appl. Chem.*, **63**, 519.
8. Hashimoto, S.-i., Yanagiya, Y., Honda, T., Harada, H. and Ikegami, S. (1992). *Tetrahedron Lett.*, **33**, 3523.
9. Hariprasad, V., Singh, G. and Tranoy, I. (1998). *Chem. Commun.*, 2129.
10. Plante, O. J., Andrade, R. B. and Seeberger, P.H. (1999). *Org. Lett.*, **1**, 211.
11. Kondo, H., Aoki, S., Ichikawa, Y., Halcomb, R. L., Ritzen, H. and Wong, C.-H. (1994). *J. Org. Chem.*, **59**, 864.
12. Müller, T., Schneider, R. and Schmidt, R. R. (1994). *Tetrahedron Lett.*, **35**, 4763.
13. Hashimoto, S.-i., Sano, A., Umeo, K., Nakajima, M. and Ikegami, S. (1995). *Chem. Pharm. Bull.*, **43**, 2267.
14. Tanaka, H., Sakamoto, H., Sano, A., Nakamura, S., Nakajima, M. and Hashimoto, S. (1999). *Chem. Commun.*, 1259.
15. Marra, A., Esnault, J., Veyrières, A. and Sinaÿ, P. (1992). *J. Am. Chem. Soc.*, **114**, 6354.
16. Barrett, A. G. M., Bezuidenhoudt, B. C. B., Gasiecki, A. F., Howell, A. R. and Russell, M. A. (1989). *J. Am. Chem. Soc.*, **111**, 1392.
17. Chenault, H. K. and Castro, A. (1994). *Tetrahedron Lett.*, **35**, 9145.
18. Boons, G.-J. and Isles, S. (1994). *Tetrahedron Lett.*, **35**, 3593.
19. Hanessian, S. ed. (1997). *Preparative Carbohydrate Chemistry*, Marcel Dekker, New York, pp. 381–466.

20. Hanessian, S., Bacquet, C. and Lehong, N. (1980). *Carbohydr. Res.*, **80**, C17.
21. Hanessian, S., Lu, P.-P. and Ishida, H. (1998). *J. Am. Chem. Soc.*, **120**, 13296.
22. Furukawa, H., Koide, K., Takao, K.-i. and Kobayashi, S. (1998). *Chem. Pharm. Bull.*, **46**, 1244.
23. Hanessian, S. and Pernet, A. G. (1976). *Adv. Carbohydr. Chem. Biochem.*, **33**, 111.
24. Postema, M. H. D. (1992). *Tetrahedron*, **48**, 8545.
25. Jaramillo, C. and Knapp, S. (1994). *Synthesis*, 1.
26. Postema, M. H. D. (1995). C-*Glycoside Synthesis*, CRC Press, Boca Raton.
27. Levy, D. E. and Tang, C. (1995). *The Chemistry of* C-*Glycosides*, Pergamon, Oxford.
28. Bertozzi, C. and Bednarski, M. (1996). Synthesis of *C*-glycosides; stable mimics of *O*-glycosidic linkages, in *Modern Methods in Carbohydrate Synthesis*, Khan, S. H. and O'Neill, R. A. eds, Harwood Academic, Netherlands, p. 316.
29. Beau, J.-M. and Gallagher, T. (1997). *Topics Curr. Chem.*, **187**, 1.
30. Nicotra, F. (1997). *Topics Curr. Chem.*, **187**, 55.
31. Du, Y., Linhardt, R. J. and Vlahov, I. R. (1998). *Tetrahedron*, **54**, 9913.
32. Sezaki, M., Kondo, S., Maeda, K., Umezawa, H. and Ohno, M. (1970). *Tetrahedron*, **26**, 5171.
33. Kishi, Y. (1998). *Pure Appl. Chem.*, **70**, 339.
34. Espinosa, J.-F., Bruix, M., Jarreton, O., Skrydstrup, T., Beau, J.-M. and Jiménez-Barbero, J. (1999). *Chem. Eur. J.*, **5**, 442.
35. Best, W. M., Ferro, V., Harle, J., Stick, R. V. and Tilbrook, D. M. G. (1997). *Aust. J. Chem.*, **50**, 463.
36. Xin, Y.-C., Zhang, Y.-M., Mallet, J.-M., Glaudemans, C. P. J. and Sinaÿ, P. (1999). *Eur. J. Org. Chem.*, 471.
37. Evans, D. A., Trotter, B. W. and Côté, B. (1998). *Tetrahedron Lett.*, **39**, 1709.
38. Baumberger, F. and Vasella, A. (1983). *Helv. Chim. Acta*, **66**, 2210.
39. Urban, D., Skrydstrup, T. and Beau, J.-M. (1998). *J. Org. Chem.*, **63**, 2507.
40. Belica, P. S. and Franck, R. W. (1998). *Tetrahedron Lett.*, **39**, 8225.
41. Jarreton, O., Skrydstrup, T., Espinosa, J.-F., Jiménez-Barbero, J. and Beau, J.-M. (1999). *Chem. Eur. J.*, **5**, 430.
42. Hoffmann, M., Burkhart, F., Hessler, G. and Kessler, H. (1996). *Helv. Chim. Acta*, **79**, 1519.
43. Beau, J.-M. and Sinaÿ, P. (1985). *Tetrahedron Lett.*, **26**, 6185, 6189, 6193.
44. Lesimple, P., Beau, J.-M., Jaurand, G. and Sinaÿ, P. (1986). *Tetrahedron Lett.*, **27**, 6201.
45. Togo, H., He, W., Waki, Y. and Yokoyama, M. (1998). *Synlett*, 700.
46. Giese, B. (1986). *Radicals in Organic Synthesis: Formation of Carbon–carbon Bonds*, Pergamon Press, Oxford.
47. Curran, D. P., Porter, N. A. and Giese, B. (1996). *Stereochemistry of Radical Reactions*, VCH, Weinheim.
48. Giese, B. and Zeitz, H.-G. (1997). *C*-Glycosyl compounds from free radical reactions, in *Preparative Carbohydrate Chemistry*, Hanessian, S. ed., Marcel Dekker, New York, p. 507.
49. Paulsen, H. and Matschulat, P. (1991). *Liebigs Ann. Chem.*, 487.
50. Nagy, J. O. and Bednarski, M. D. (1991). *Tetrahedron Lett.*, **32**, 3953.
51. Sinaÿ, P. (1997, 1998, 1998). *Pure Appl. Chem.*, **69**, 459; **70**, 407; **70**, 1495.
52. Rubinstenn, G., Mallet, J.-M. and Sinaÿ, P. (1998). *Tetrahedron Lett.*, **39**, 3697.
53. Vasella, A. (1991, 1993). *Pure Appl. Chem.*, **63**, 507; **65**, 731.
54. Curran, D. P. and Suh, Y.-G. (1987). *Carbohydr. Res.*, **171**, 161.
55. Brakta, M., Lhoste, P. and Sinou, D. (1989). *J. Org. Chem.*, **54**, 1890.
56. He, W., Togo, H., Waki, Y. and Yokoyama, M. (1998). *J. Chem. Soc., Perkin Trans. 1*, 2425.

2-Acetamido-2-Deoxy and 2-Deoxy Glycosides

2-Acetamido-2-deoxy glycosides, especially those of the D-*gluco* and D-*galacto* configuration, are common components of many biopolymers (chitin, chondroitin sulfate) and glycoconjugates (glycoproteins, proteoglycans, glyco-lipids). 2-Deoxy glycosides are often found in the oligosaccharide portion of natural products derived from plant and bacterial sources.

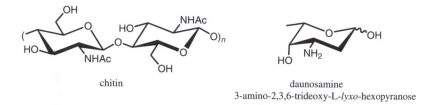

chitin

daunosamine
3-amino-2,3,6-trideoxy-L-*lyxo*-hexopyranose

These two sorts of glycosides obviously serve different functions in the host organism — the presence of an acetamido group (an amide) changes the polarity and hydrogen bonding potential of a molecule; a 2-deoxy function, among other things, makes the associated glycoside more amenable to hydrolysis. The syntheses of 2-acetamido and 2-deoxy glycosides are inexorably linked for somewhat different reasons. Whereas a 2-acetamido group is capable of participating in events occurring at the anomeric carbon (even to the extent of forming a somewhat unreactive oxazoline), the absence of an oxygen functionality at C2 gives rise to problems in the stereoselectivity of formation of the glycosidic linkage.

2-Acetamido-2-Deoxy Glycosides[1, 2]

The most obvious approach to the synthesis of 2-acetamido-2-deoxy glycosides would be to utilize a glycosyl donor with the acetamido group already in place. Indeed, this procedure works well for the synthesis of 1,2-*trans*-glycosides derived from reasonably reactive alcohols. The intermediate oxazoline is rather stable and, naturally, dominates the stereoselectivity of the event:[3]

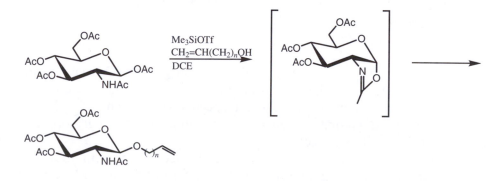

Glycosyl chlorides and glycosyl trichloroacetimidates may also be used as the donor;[4] on occasion, the oxazoline may first be prepared separately and then used in the subsequent glycosidation. The trichloroethoxycarbonyl group has proven useful for the protection of the primary amine during the glycosidation step.[5]

Much better results are obtained when *N,N*-diacyl glycosyl donors are used — the only drawback is that a deprotection (and sometimes subsequent acetylation) step is now necessary. The most common diacyl donors are based on phthalimide:[4,6–8]

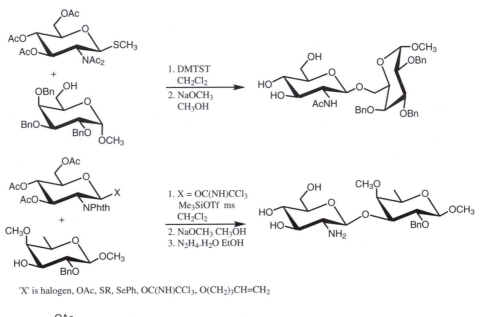

'X' is halogen, OAc, SR, SePh, OC(NH)CCl₃, O(CH₂)₃CH=CH₂

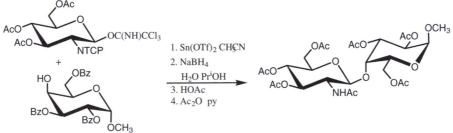

Again, the glycosidations generally produce only the 1,2-*trans* product. Although participation by the diacetylamino group at the anomeric centre is expected, it is a moot point whether the same occurs with the phthalimido group (the planar, aromatic system is generally orthogonal to the D-glycopyranose ring and thus blocks the α-face).

2-Azido-2-deoxy glycosyl donors are easily accessible from the appropriate glycal by either "azido nitration" or "azido selenation" and offer convenient methods for the stereoselective synthesis of 2-acetamido-2-deoxy-

D-glycopyranosides:[2,4,9,10]

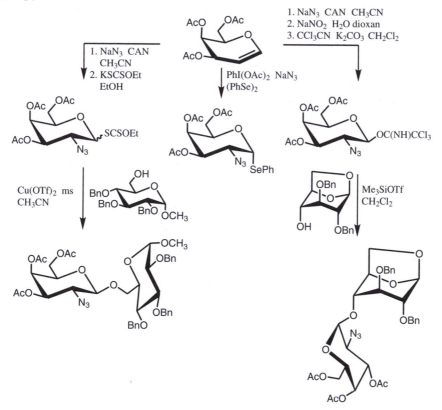

Other 2-azido-2-deoxy glycosyl donors have been used for related syntheses.[11,12]

Glycals have also been used for the synthesis of 2-acetamido-2-deoxy-β-D-glycopyranosides, or their direct precursors:[13,14]

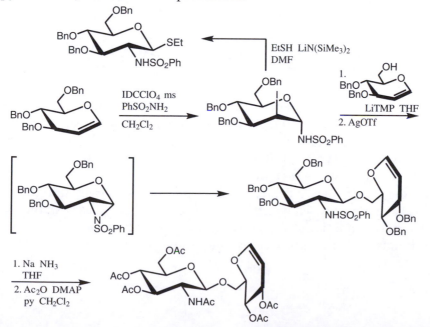

Finally, as with the synthesis of β-D-mannopyranosides, special methods had to be developed for their 2-acetamido-2-deoxy counterparts; one of the best appears to involve the stereoselective reduction of a 2-ulose oxime:[15]

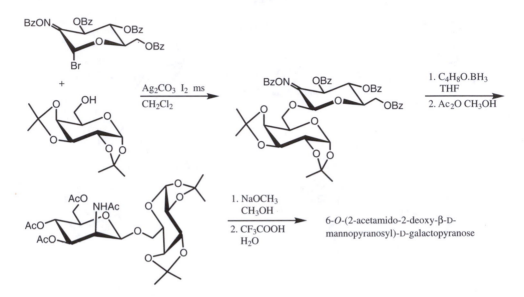

2-Deoxy Glycosides[16,17]

As mentioned previously, any method of synthesis of a 2-deoxy glycoside that simply delivers a hydrogen atom to C2 of a glycal has the potential for poor stereoselectivity:[18]

However, in spite of the absence of a participating group at C2, several 2-deoxy glycosyl donors, including thioglycosides,[19] phosphonodithioates,[20,21] trichloroacetimidates,[22] fluorides,[23] tetrazoles and phosphites,[24] have been used successfully for the synthesis of a variety of glycosides — the main product is generally the more stable α-D-glycoside.

In order to introduce a degree of anomeric stereoselectivity into the process, methods have been devised which actually functionalize C2 as well; the group at C2 is temporary and is subsequently removed after the glycosidation step. An added bonus of this approach is that only the final product, lacking a substituent at C2, is labile towards acid:[25-30]

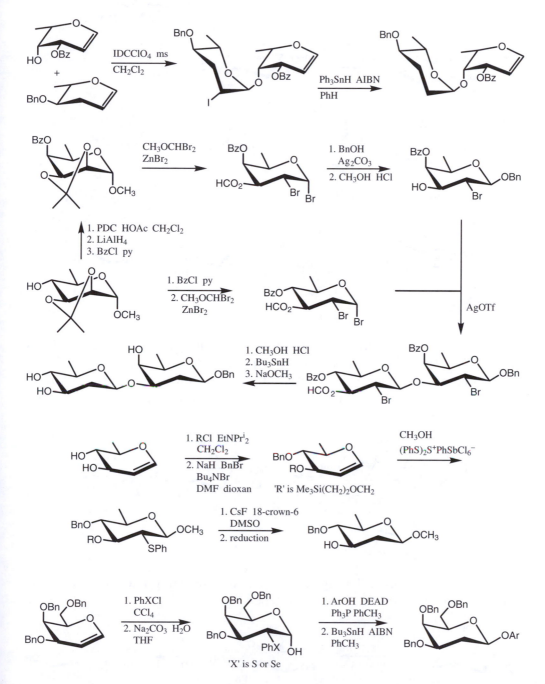

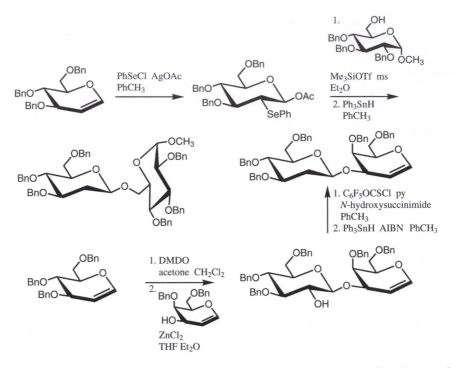

As can be seen from the above results, good methods now exist for the synthesis of 2-deoxy β-D- and α-L-glycosides; other procedures are also possible.[31,32]

Nicolaou has reported a very neat rearrangement of a thioglycoside, powered by diethylaminosulfur trifluoride and leading to a glycosyl fluoride suitable for the synthesis of both 2-deoxy α- and β-glycosides:[33]

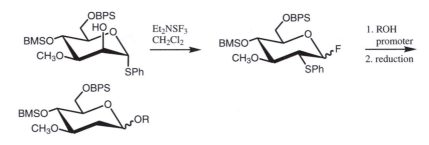

van Boom, in a process that again involves an interesting rearrangement, has devised a direct synthesis of the same type of glycoside:[34]

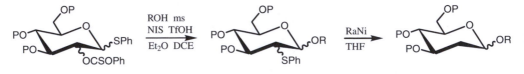

'P' is a protecting group

Other novel methods for the stereoselective synthesis of 2-deoxy glycosides continue to be announced.[35-37]

References

1. Banoub, J., Boullanger, P. and Lafont, D. (1992). *Chem. Rev.*, **92**, 1167.
2. Veeneman, G. H. (1998). Chemical synthesis of *O*-glycosides, in *Carbohydrate Chemistry*, Boons, G.-J. ed., Blackie, Edinburgh, p. 98.
3. Fairweather, J. K., Stick, R. V. and Tilbrook, D. M. G. (1998). *Aust. J. Chem.*, **51**, 471.
4. Schmidt, R. R. and Kinzy, W. (1994). *Adv. Carbohydr. Chem. Biochem.*, **50**, 21.
5. Meinjohanns, E., Meldal, M. and Bock, K. (1995). *Tetrahedron Lett.*, **36**, 9205.
6. Castro-Palomino, J. C. and Schmidt, R. R. (1995). *Tetrahedron Lett.*, **36**, 6871, 5343.
7. Silwanis, B. A., El-Sokkary, R. I., Nashed, M. A. and Paulsen, H. (1991). *J. Carbohydr. Chem.*, **10**, 1067.
8. Leung, O.-T., Douglas, S. P., Whitfield, D. M., Pang, H. Y. S. and Krepinsky, J. J. (1994). *New J. Chem.*, **18**, 349.
9. Grundler, G. and Schmidt, R. R. (1984). *Liebigs Ann. Chem.*, 1826.
10. Czernecki, S., Ayadi, E. and Randriamandimby, D. (1994). *J. Org. Chem.*, **59**, 8256.
11. Marra, A., Gauffeny, F. and Sinaÿ, P. (1991). *Tetrahedron*, **47**, 5149.
12. Ludewig, M. and Thiem, J. (1998). *Eur. J. Org. Chem.*, 1189.
13. Danishefsky, S. J. and Bilodeau, M. T. (1996). *Angew. Chem. Int. Ed. Engl.*, **35**, 1380.
14. Di Bussolo, V., Liu, J., Huffman, L. G. Jr and Gin, D. Y. (2000). *Angew. Chem. Int. Ed.*, **39**, 204.
15. Kaji, E., Lichtenthaler, F. W., Nishino, T., Yamane, A. and Zen, S. (1988). *Bull. Chem. Soc. Jpn.*, **61**, 1291.
16. Thiem, J. and Klaffke, W. (1990). *Topics Curr. Chem.*, **154**, 285.
17. Kirschning, A., Bechthold, A. F.-W. and Rohr, J. (1997). *Topics Curr. Chem.*, **188**, 1.
18. Toshima, K., Nagai, H., Ushiki, Y. and Matsumura, S. (1998). *Synlett*, 1007.
19. Nicolaou, K. C., Seitz, S. P. and Papahatjis, D. P. (1983). *J. Am. Chem. Soc.*, **105**, 2430.
20. Laupichler, L., Sajus, H. and Thiem, J. (1992). *Synthesis*, 1133.
21. Bielawska, H. and Michalska, M. (1998). *Tetrahedron Lett.*, **39**, 9761.
22. Schene, H. and Waldmann, H. (1998). *Chem. Commun.*, 2759.
23. Jünnemann, J., Lundt, I. and Thiem, J. (1991). *Liebigs Ann. Chem.*, 759.
24. Guo, Y. and Sulikowski, G. A. (1998). *J. Am. Chem. Soc.*, **120**, 1392.
25. Suzuki, K., Sulikowski, G. A., Friesen, R. W. and Danishefsky, S. J. (1990). *J. Am. Chem. Soc.*, **112**, 8895.
26. Thiem, J. and Schöttmer, B. (1987). *Angew. Chem. Int. Ed. Engl.*, **26**, 555.
27. Franck, R. W. and Kaila, N. (1993). *Carbohydr. Res.*, **239**, 71.
28. Roush, W. R. and Lin, X.-F. (1991). *J. Org. Chem.*, **56**, 5740.
29. Perez, M. and Beau, J.-M. (1989). *Tetrahedron Lett.*, **30**, 75.
30. Gervay, J. and Danishefsky, S. (1991). *J. Org. Chem.*, **56**, 5448.
31. Trumtel, M., Veyrières, A. and Sinaÿ, P. (1989). *Tetrahedron Lett.*, **30**, 2529.
32. Tavecchia, P., Trumtel, M., Veyrières, A. and Sinaÿ, P. (1989). *Tetrahedron Lett.*, **30**, 2533.
33. Nicolaou, K. C., Ladduwahetty, T., Randall, J. L. and Chucholowski, A. (1986). *J. Am. Chem. Soc.*, **108**, 2466.

34. Zuurmond, H. M., van der Klein, P. A. M., van der Marel, G. A. and van Boom, J. H. (1993). *Tetrahedron*, **49**, 6501.
35. Roush, W. R., Gung, B. W. and Bennett, C. E. (1999). *Org. Lett.*, **1**, 891.
36. Roush, W. R., Narayan, S., Bennett, C. E. and Briner, K. (1999). *Org. Lett.*, **1**, 895.
37. Roush, W. R. and Narayan, S. (1999). *Org. Lett.*, **1**, 899.

Chapter 9

Oligosaccharide Synthesis

Strategies in Oligosaccharide Synthesis

I did it, my way
    Frank Sinatra

If you were to enquire as to the best way of constructing a particular glycosidic linkage in an oligosaccharide of interest, the answer would often depend on the person being asked:

Hans Paulsen	"use a glycosyl halide"
Ray Lemieux	"halide-catalysis, of course"
K. C. Nicolaou or	"a glycosyl fluoride"
Terukai Mukaiyama	
Richard Schmidt	"trichloroacetimidates!"
Per Garegg	"a thioglycoside"
Dan Kahne	"a glycosyl sulfoxide"
Sam Danishefsky	"glycal assembly"

All offer good advice. In fact, there is no one method of glycosidation that is general and reliable enough to guarantee success. The situation is epitomized in the much quoted lines of Hans Paulsen:[2]

> *Each oligosaccharide synthesis remains an independent problem, whose resolution requires considerable systematic research and a good deal of know-how. There are no universal reaction conditions for oligosaccharide syntheses.*

Every dog has its day
Anon

From our previous discussions, we now have an appreciation of the various factors that influence the glycosidic bond, as well as a summary of the established and the newer methods for its construction. What is lacking in our knowledge is the *philosophy* behind oligosaccharide synthesis.

Linear Syntheses

A *linear synthesis* is one in which a monosaccharide, by the successive addition of other monosaccharides, is transformed into the desired oligosaccharide:[a]

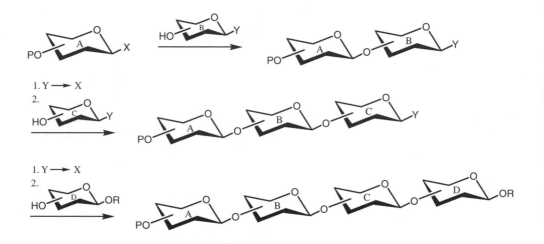

A protected glycosyl donor (A, X activated) is treated with an acceptor alcohol (B, Y inactive) to generate a disaccharide (AB, Y inactive). After the conversion of Y into X, the process is repeated until the final target oligosaccharide is reached.

Several modern versions of this linear approach exist, some of which involve "*two-stage activation*" methods:[3–6]

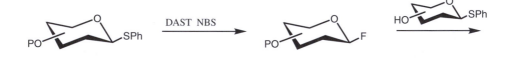

---

[a] In order to simplify the diagramatic presentations, no stereochemistry is implied at the anomeric carbon of any generalized structure.

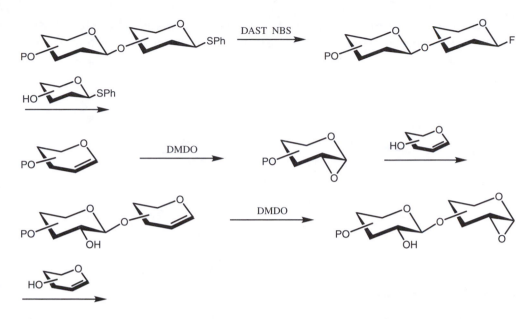

The success of many of these methods obviously relies on the *selective activation* of one glycosyl donor over another (here, fluoride over thioglycoside and epoxide over alkene) and there are countless other examples (for example, bromide over thioglycoside[7] and selenoglycoside over thioglycoside[8]). The "latent/active" concept discussed earlier is another variant here, as is the concept of "orthogonality".[1]

Another useful approach involves the *chemoselective activation* of one glycosyl donor over another and is very much connected with the "armed/disarmed" concept discussed earlier:

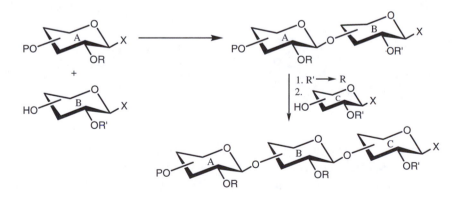

Here, the glycosides (A) and (B) possess the same aglycon but act as donor and acceptor, respectively, because of the activating nature of R (A) and the deactivating nature of **R'** (**B**). The phenomenon was first noticed with 4-pentenyl glycosides[9] and was later extended to thioglycosides and related molecules.[1,10]

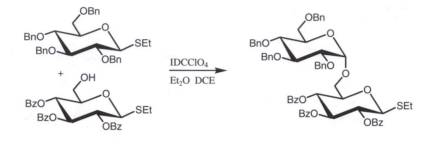

Convergent Syntheses

A *convergent synthesis* assembles the oligosaccharide from smaller, preformed components:

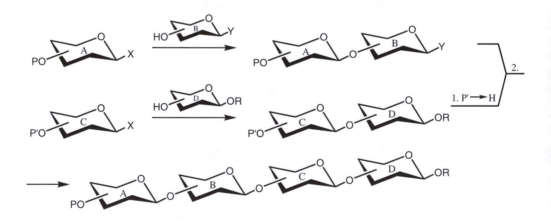

The convergent approach, often dubbed a "block synthesis" (for obvious reasons), has several advantages over its linear counterpart. The smaller components can often be obtained from naturally occurring disaccharides (lactose, cellobiose), an expensive monosaccharide can be introduced late in the synthetic sequence and, generally, there are fewer manipulations that need to be performed on the growing oligosaccharide chain. In general, the concepts of two-stage and selective/chemoselective activation, "latent/active" and ortho-gonality lend themselves very nicely to convergent syntheses.[1,11]

Two-directional Syntheses

Boons has recently reported a highly convergent approach to oligosaccharide synthesis, whereby a monosaccharide derivative is constructed such that it is capable of acting as both a glycosyl donor and a glycosyl acceptor:[12]

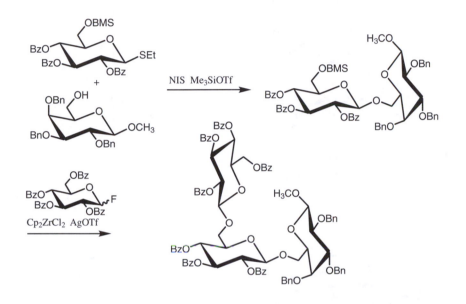

This innovative method, which builds on many of the principles established for glycoside synthesis, has been applied to the synthesis of a pentasaccharide that is associated with the hyperacute rejection response in xenotransplantation from pig to man.[13]

"One-pot" Syntheses

The ultimate goal in oligosaccharide synthesis is to be able to assemble the molecule in question in "one-pot", *viz.* by the step-wise addition of building blocks to the growing chain, with no need for manipulation of protecting groups and anomeric activating groups and with complete stereochemical control. Such an achievement would rival the successes of polymer-based syntheses (to be discussed next) and notable strides have been made after the original suggestion of the concept by Takahashi:[11,14–17]

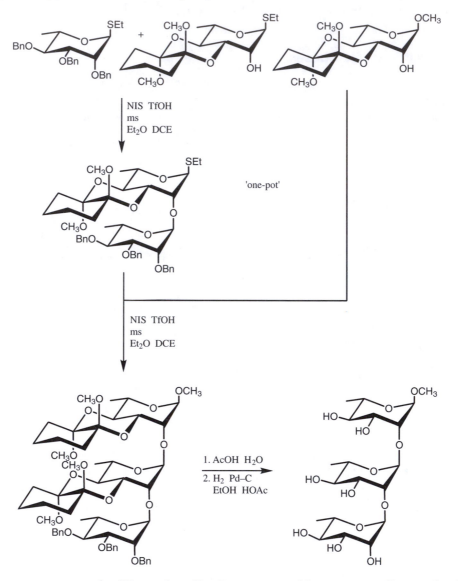

Two recent papers by Wong describe "programmable, one-pot oligosaccharide synthesis".[18,19]

To complete this section, a selection of recent syntheses of oligosaccharides is given.[20–31] Barresi and Hindsgaul have developed a "searchable table" of glycosidic linkages.[32, b]

---

[b] http://www.chem.ualberta.ca/~glyco/misc/storage/glycosyl_top.html

References

1. Boons, G.-J. (1996). *Tetrahedron*, **52**, 1095.
2. Paulsen, H. (1982). *Angew. Chem. Int. Ed. Engl.*, **21**, 155.
3. Nicolaou, K. C., Caulfield, T. J. and Groneberg, R. D. (1991). *Pure Appl. Chem.*, **63**, 555.
4. Sliedregt, L. A. J. M., van der Marel, G. A. and van Boom, J. H. (1994). *Tetrahedron Lett.*, **35**, 4015.
5. Friesen, R. W. and Danishefsky, S. J. (1989). *J. Am. Chem. Soc.*, **111**, 6656.
6. Halcomb, R. L. and Danishefsky, S. J. (1989). *J. Am. Chem. Soc.*, **111**, 6661.
7. Fügedi, P., Birberg, W., Garegg, P. J. and Pilotti, A. (1987). *Carbohydr. Res.*, **164**, 297.
8. Mehta, S. and Pinto, B. M. (1993). *J. Org. Chem.*, **58**, 3269.
9. Mootoo, D. R., Konradsson, P., Udodong, U. and Fraser-Reid, B. (1998). *J. Am. Chem. Soc.*, **110**, 5583.
10. Veeneman, G. H., van Leeuwen, S. H. and van Boom, J. H. (1990). *Tetrahedron Lett.*, **31**, 1331.
11. Boons, G.-J. and Zhu, T. (1997). *Synlett*, 809.
12. Zhu, T. and Boons, G.-J. (1998). *Tetrahedron Lett.*, **39**, 2187.
13. Zhu, T. and Boons, G.-J. (1998). *J. Chem. Soc., Perkin Trans. 1*, 857.
14. Yamada, H., Harada, T., Miyazaki, H. and Takahashi, T. (1994). *Tetrahedron Lett.*, **35**, 3979.
15. Yamada, H., Harada, T. and Takahashi, T. (1994). *J. Am. Chem. Soc.*, **116**, 7919.
16. Raghavan, S. and Kahne, D. (1993). *J. Am. Chem. Soc.*, **115**, 1580.
17. Ley, S. V. and Priepke, H. W. M. (1994). *Angew. Chem. Int. Ed. Engl.*, **33**, 2292.
18. Zhang, Z., Ollmann, I. R., Ye, X.-S., Wischnat, R., Baasov, T. and Wong, C.-H. (1999). *J. Am. Chem. Soc.*, **121**, 734.
19. Ye, X.-S. and Wong, C.-H. (2000). *J. Org. Chem.*, **65**, 2410.
20. Rana, S. S., Vig, R. and Matta, K. L. (1982–83). *J. Carbohydr. Chem.*, **1**, 261.
21. Paulsen, H. (1990). *Angew. Chem. Int. Ed. Engl.*, **29**, 823.
22. Dan, A., Ito, Y. and Ogawa, T. (1995). *J. Org. Chem.*, **60**, 4680.
23. Demchenko, A. and Boons, G.-J. (1997). *Tetrahedron Lett.*, **38**, 1629.
24. Nilsson, M. and Norberg, T. (1998). *J. Chem. Soc., Perkin Trans. 1*, 1699.
25. Kwon, O. and Danishefsky, S. J. (1998). *J. Am. Chem. Soc.*, **120**, 1588.
26. Crich, D. and Dai, Z. (1998). *Tetrahedron Lett.*, **39**, 1681.
27. Weiler, S. and Schmidt, R. R. (1998). *Tetrahedron Lett.*, **39**, 2299.
28. Schmidt, R. R. (1998). *Pure Appl. Chem.*, **70**, 397.
29. Garegg, P. J., Konradsson, P., Lezdins, D., Oscarson, S., Ruda, K. and Öhberg, L. (1998). *Pure Appl. Chem.*, **70**, 293.
30. Baeschlin, D. K., Chaperon, A. R., Green, L. G., Hahn, M. G., Ince, S. J. and Ley, S. V. (2000). *Chem. Eur. J.*, **6**, 172.
31. Düffels, A. and Ley, S. V. (1999). *J. Chem. Soc., Perkin Trans. 1*, 375.
32. Barresi, F. and Hindsgaul, O. (1995). *J. Carbohydr. Chem.*, **14**, 1043.

Polymer-supported Synthesis of Oligosaccharides[1-8]

Ever since the invention of the "solid phase" method for the synthesis of oligopeptides by Merrifield (which, ultimately, attracted the award of a Nobel Prize in 1984[9]), and the similar successes enjoyed with oligonucleotides later, carbohydrate chemists have dreamt of utilizing the method for the synthesis of oligosaccharides. However, whereas a growing peptide chain on a resin bead presents, after minimal protecting group manoeuvres, a single primary amine (except for proline) of predictable reactivity for coupling to a likewise predictable carboxylic acid, such is not the case with carbohydrates. Whether a monosaccharide is attached to a resin bead *via* the anomeric or a non-anomeric hydroxyl group, the subsequent construction of a glycosidic linkage is plagued by all of the common factors — protecting groups, regioselectivity, stereoselectivity, solvent, donor, promotor, etc. In addition, whereas the twenty or so common amino acids are all available with the amino and side-chain functional groups protected in any number of reliable ways, such is not the case with monosaccharides; there is, as yet, not even an orthogonally protected form of D-glucose that could be viewed as a general glycosyl donor.[10] However, enormous advances in the use of polymers for the synthesis of oligosaccharides have been made over the last decade and a broad overview of the area will be given here.

Types of Polymers[11]

Of the *insoluble polymers*, which are available as discrete beads, the original (Merrifield) polystyrene (PS) is still used, usually chloromethylated to enable attachment of the sugar. "Wang resin"[12] offers a primary alcohol for attachment and this can be modified to form an aldehyde;[13] "TentaGel" has a polystyrene core that is functionalizd with poly(ethylene glycol).[14,15] Whereas all of the above resins need to "swell" to achieve good "loadings", such is not the case with "controlled-pore glass" (CPG) — the highly hydroxylated surface is ideal for functionalization by carbohydrates.[16,17] Most of the resins are commercially available with a variety of different functional groups, for example, Merrifield resin with Cl, OH or NH_2 and TentaGel with Br, OH, NH_2, SH or COOH.

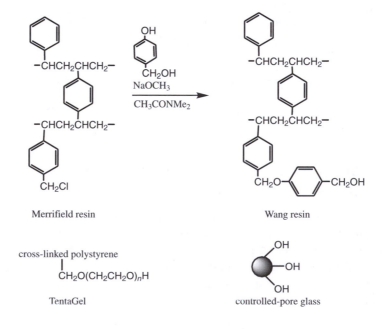

Soluble polymers offer the advantage of being able to conduct the desired chemical reactions in solution, yet display their versatility as insoluble solids during the purification step. The most common soluble polymer that is used in the synthesis of oligosaccharides is *O*-methyl poly(ethylene glycol) (MPEG):[18,19]

$$CH_3O(CH_2CH_2O)_nH$$

Linkers

One sugar of the desired oligosaccharide must, naturally, be first attached to the polymer. This is usually done employing a *linker* that, although stable to all of the synthetic manipulations that will be carried out on the polymer, must be easily broken to liberate the final product. Because of the lability of the glycosidic linkage towards acid, the useful linkers employed are esters,[15,17] benzylic ethers,[20] photolabile groups,[1,21,22] silyl ethers[23] and silanes,[24] all removed by mild and selective reagents.

The Attachment of the Sugar to the Polymer

There are obviously just two discrete ways to attach the first monosaccharide to the polymer, either through the anomeric centre (a glycosidic linkage) or utilizing one of the other hydroxyl groups on the (pyranose) ring:

For the attachment through a glycosidic linkage, a subsequent deprotection step liberates the hydroxyl group of interest and another glycosidation can be performed to give a disaccharide now attached to the linked polymer. For the non-anomeric linkage, an acceptor and a promotor are added to the polymer-bound glycosyl donor and an attached disaccharide again results. There are arguments for and against either approach but, in these days of active glycosyl donors, both methods have yielded spectacular results.

The Glycosyl Donors Used

In line with the success of oligosaccharide synthesis in solution, the common glycosyl donors in polymer-supported syntheses are trichloroacetimidates, pentenyl glycosides, glycosyl sulfoxides, thioglycosides and glycals.

Insoluble *versus* Soluble Polymers

Although much of the early work on the polymer-supported synthesis of oligosaccharides was carried out on insoluble resins, it is probably safe to say that the modern trend is towards the use of soluble polymers.

In the syntheses using insoluble polymers, a commercial resin (purchased as beads) is swollen in a solvent and the chosen linker attached. Next, the first monosaccharide is joined to the linker (either through the reducing or non-reducing end), with complete reaction ensured by the use of an excess of the reagent(s). Following a wash step, the cycle is repeated until the desired oligosaccharide has been assembled. The deprotection of the many hydroxyl groups generally precedes the liberation of the oligosaccharide from the resin.

For the soluble polymer approach, a commercial product is again available which is again attached to a suitable linker, utilizing any solvent apart from ethers. The first monosaccharide is then attached to the linker, *via* the anomeric or a non-anomeric hydroxyl group (this choice depends on the nature of the linker).[5,19] In the former case, the deprotection of the desired hydroxyl group, followed by treatment with an excess of the next glycosyl donor and promotor, gives the disaccharide attached to the polymer, still in solution. The addition of an ether (usually *tert*-butyl methyl ether) precipitates this product which may be collected and washed. Repetition of the cycle yields, after deprotection and cleavage from the resin, the desired oligosaccharide.

Some Examples

Trichloroacetimidates:[16, 25]

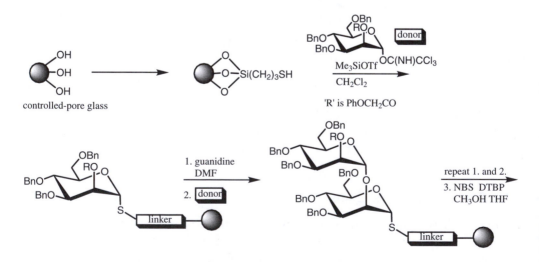

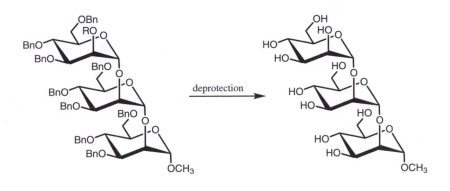

For such polymer-supported syntheses to be successful, it has been shown that the choice of polymer is important,[17] that more active promotors (for example, dibutylboron triflate with trichloroacetimidates[26]) may be needed to counter the often sluggish glycosylation steps, that multiple exposure to reagents is often required,[15] and that "capping" of unreacted residues at each stage of the synthesis helps in improving the purity of the final product.[15,27] Somewhat surprisingly, there is often a marked improvement in the anomeric stereo-selectivity for glycosidations performed on a polymer support over those conducted in normal solution.[28]

Pentenyl glycosides:[22]

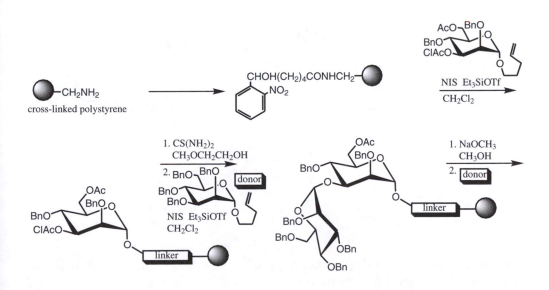

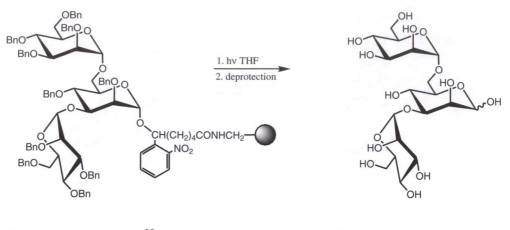

Glycosyl sulfoxides:[29]

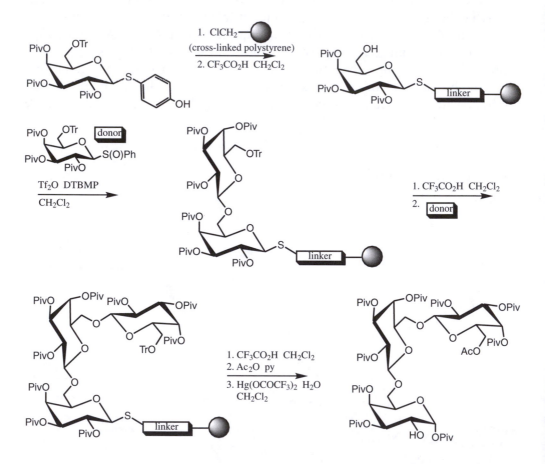

Thioglycosides: Several polymer-supported syntheses utilizing thioglyco-sides as the donor have been reported. Two of interest utilize *O*-methyl poly(ethylene glycol) with a succinoyl linker,[27] and Merrifield resin, again with a photolabile linker:[30]

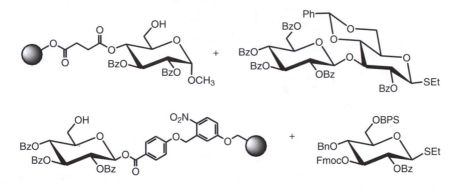

Glycals:[31] Danishefsky, in line with the success of the "glycal assembly" approach for the synthesis of oligosaccharides in solution, has applied the method to a solid support, generally attaching the first monosaccharide as a glycosyl *donor* — one of the most impressive outcomes is a synthesis of the Lewis[b] blood group determinant glycal:[31,32]

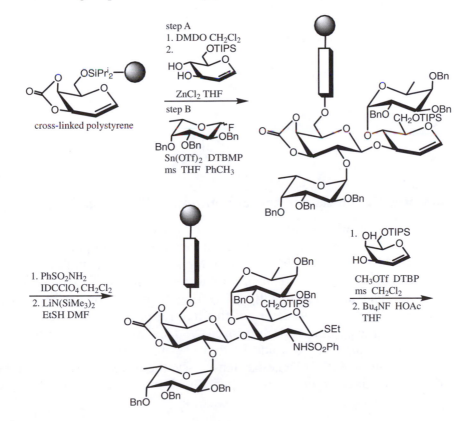

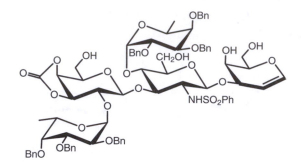

In his work, Danishefsky has never shied away from converting his beloved glycal, when necessary, into a more appropriate thioglycoside.[31,33]

Finally, Ito and Ogawa have made significant and recent contributions to polymer-supported oligosaccharide synthesis. Apart from the application of the orthogonal glycosylation strategy[34] and intramolecular aglycon delivery,[35] these two collaborators have introduced the concept of a "hydrophobic tag" which allows for the convenient purification (reversed-phase silica gel) of the oligosaccharide after detachment from the polymer:

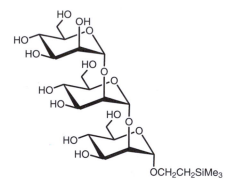

Combinatorial Synthesis and the Generation of "Libraries"

The last decade has witnessed a complete change in the philosophy for the discovery of compounds that, eventually, may lead to a novel pharmaceutical. Traditionally, chemists have worked with botanists, zoologists and even native people to select plants and organisms that, upon chemical extraction and analysis, may yield a chemical with interesting pharmacological properties. Tinkering with the chemical structure of this natural product has often given a better pharmaceutical, one with improved and/or a broader spectrum of activity — a classical example would be that of penicillin.

In these traditional approaches, extensive synthetic effort was invariably required — many have experienced the heartbreak of a long synthesis, only to find that the target molecule exhibited anything but the desired properties. The computer-assisted design of target molecules has improved things somewhat but is generally only of use when a deal is known about the mechanism of action of a disease-causing organism.[36]

A novel approach to "drug discovery" was needed and so was born *"combinatorial synthesis"*[1,37-42] — the generation of a "library" of closely related structures by (solid) polymer-supported synthesis:

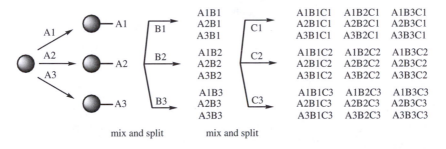

A1B1		A1B1C1 A1B2C1 A1B3C1
A2B1		A2B1C1 A2B2C1 A2B3C1
A3B1		A3B1C1 A3B2C1 A3B3C1

mix and split   mix and split

In the "mix and split" approach illustrated above, a functionalized bead sample is treated separately with three different, but related, reagents (A1–A3) to generate three newly functionalized bead samples. These three bead samples are then mixed and subsequently split into three before being separately treated again with three new reagents (B1–B3). The process is repeated at will and the final library consists of resin beads, to which are attached multiple copies of a unique chemical. Screening of this library for some sort of biological or pharmacological activity is followed by isolation of the active bead and identification of the unique chemical in question — this is where the fun and games start!

Although it is often a straightforward matter to generate a library of many thousands of different compounds, the screening and structure identification steps present huge problems[43-45] — no one has put it better than Clark Still:[46]

> As for encoding in particular, it seems to complete one of the most powerful of combinatorial methodologies: split-and-pool synthesis and on-bead property screening. Thus, split-and-pool synthesis provides access to large libraries conveniently, on-bead screening allows the efficient selection of library members having a property of interest, and encoding provides a straightforward path to the structures of selected members. These three methods work exceptionally well together and provide a complete methodology for combinatorial exploration of chemical design problems.

Owing to the polyvalent nature of monosaccharides and to the general difficulties associated with anomeric stereocontrol, it was not surprising that it took until 1995 for Hindsgaul to announce the first preparation of a library of carbohydrates (trisaccharides), performed in solution without the aid of a polymer support.[47] This result was soon followed by a report from Boons, detailing the preparation of a library of trisaccharides with no attempt to control the anomeric stereoselectivity (this simple and expedient operation substantially increases the complexity of the library).[48]

Polymer-supported syntheses were soon to follow, with one of the most notable involving the use of glycosyl sulfoxides and an ingenious screening/encoding system that allowed the identification and structure determination of an "active" molecule.[49-51]

Many improvements in the construction of carbohydrate libraries, both on and off polymer supports, continue to be announced.[52-59]

References

1. Osborn, H. M. I. and Khan, T. H. (1999). *Tetrahedron*, **55**, 1807.
2. Brown, A. R., Hermkens, P. H. H., Ottenheijm, H. C. J. and Rees, D. C. (1998). *Synlett*, 817.
3. Danishefsky, S. J., McClure, K. F., Randolph, J. T. and Ruggeri, R. B. (1993). *Science*, **260**, 1307.
4. Seitz, O. (1998). *Angew. Chem. Int. Ed.*, **37**, 3109.
5. Früchtel, J. S. and Jung, G. (1996). *Angew. Chem. Int. Ed. Engl.*, **35**, 17.
6. Hermkens, P. H. H., Ottenheijm, H. C. J. and Rees, D. (1996). *Tetrahedron*, **52**, 4527.
7. Brown, R. C. D. (1998). *J. Chem. Soc., Perkin Trans. 1*, 3293.
8. Ito, Y. and Manabe, S. (1998). *Curr. Opin. Chem. Biol.*, **2**, 701.
9. Merrifield, R. B. (1985). *Angew. Chem. Int. Ed. Engl.*, **24**, 799.
10. Zhu, T. and Boons, G.-J. (2000). *Tetrahedron*: Asymmetry, **11**, 199.
11. Sherrington, D. C. (1998). *Chem. Commun.*, 2275.
12. Wang, S.-S. (1973). *J. Am. Chem. Soc.*, **95**, 1328.
13. Hanessian, S. and Huynh, H. K. (1999). *Synlett*, 102.
14. Bayer, E. (1991). *Angew. Chem. Int. Ed. Engl.*, **30**, 113.
15. Adinolfi, M., Barone, G., De Napoli, L., Iadonisi, A. and Piccialli, G. (1996). *Tetrahedron Lett.*, **37**, 5007.
16. Heckel, A., Mross, E., Jung, K.-H., Rademann, J. and Schmidt, R. R. (1998). *Synlett*, 171.
17. Adinolfi, M., Barone, G., De Napoli, L., Iadonisi, A. and Piccialli, G. (1998). *Tetrahedron Lett.*, **39**, 1953.
18. Gravert, D. J. and Janda, K. D. (1997). *Chem. Rev.*, **97**, 489.
19. Krepinsky, J. J. (1996). Advances in polymer-supported solution synthesis of oligosaccharides, in *Modern Methods in Carbohydrate Synthesis*, Khan, S. H. and O'Neill, R. A. eds, Harwood Academic, Netherlands, p. 194.

20. Hodosi, G. and Krepinsky, J. J. (1996). *Synlett*, 159.
21. Nicolaou, K. C., Winssinger, N., Pastor, J. and DeRoose, F. (1997). *J. Am. Chem. Soc.*, **119**, 449.
22. Rodebaugh, R., Fraser-Reid, B. and Geysen, H. M. (1997). *Tetrahedron Lett.*, **38**, 7653.
23. Doi, T., Sugiki, M., Yamada, H., Takahashi, T. and Porco, J. A., Jr. (1999). *Tetrahedron Lett.*, **40**, 2141.
24. Weigelt, D. and Magnusson, G. (1998). *Tetrahedron Lett.*, **39**, 2839.
25. Rademann, J., Geyer, A. and Schmidt, R. R. (1998). *Angew. Chem. Int. Ed.*, **37**, 1241.
26. Wang, Z.-G., Douglas, S. P. and Krepinsky, J. J. (1996). *Tetrahedron Lett.*, **37**, 6985.
27. Verduyn, R., van der Klein, P. A. M., Douwes, M., van der Marel, G. A. and van Boom, J. H. (1993). *Recl. Trav. Chim. Pays-Bas*, **112**, 464.
28. Hunt, J. A. and Roush, W. R. (1996). *J. Am. Chem. Soc.*, **118**, 9998.
29. Yan, L., Taylor, C. M., Goodnow, R. Jr and Kahne, D. (1994). *J. Am. Chem. Soc.*, **116**, 6953.
30. Nicolaou, K. C., Watanabe, N., Li, J., Pastor, J. and Winssinger, N. (1998). *Angew. Chem. Int. Ed.*, **37**, 1559.
31. Seeberger, P. H. and Danishefsky, S. J. (1998). *Acc. Chem. Res.*, **31**, 685.
32. Zheng, C., Seeberger, P. H. and Danishefsky, S. J. (1998). *Angew. Chem. Int. Ed.*, **37**, 786.
33. Zheng, C., Seeberger, P. H. and Danishefsky, S. J. (1998). *J. Org. Chem.*, **63**, 1126.
34. Ito, Y., Kanie, O. and Ogawa, T. (1996). *Angew. Chem. Int. Ed. Engl.*, **35**, 2510.
35. Ito, Y. and Ogawa, T. (1997). *J. Am. Chem. Soc.*, **119**, 5562.
36. Brown, R. D. and Newsam, J. M. (1998). *Chem. Ind.*, 785.
37. Terrett, N. K., Gardner, M., Gordon, D. W., Kobylecki, R. J. and Steele, J. (1995). *Tetrahedron*, **51**, 8135.
38. Lowe, G. (1995). *Chem. Soc. Rev.*, **24**, 309.
39. Thompson, L. A. and Ellman, J. A. (1996). *Chem. Rev.*, **96**, 555.
40. Balkenhohl, F., von dem Bussche-Hünnefeld, C., Lansky, A. and Zechel, C. (1996). *Angew. Chem. Int. Ed. Engl.*, **35**, 2288.
41. Tapolczay, D. J., Kobylecki, R. J., Payne, L. J. and Hall, B. (1998). *Chem. Ind.*, 772.
42. Sofia, M. J. (1996). *Drug Discov. Today*, **1**, 27.
43. Asaro, M. F. and Wilson, R. B., Jr. (1998). *Chem. Ind.*, 777.
44. Nicolaou, K. C., Xiao, X.-Y., Parandoosh, Z., Senyei, A. and Nova, M. P. (1995). *Angew. Chem. Int. Ed. Engl.*, **34**, 2289.
45. Moran, E. J., Sarshar, S., Cargill, J. F., Shahbaz, M. M., Lio, A., Mjalli, A. M. M. and Armstrong, R. W. (1995). *J. Am. Chem. Soc.*, **117**, 10787.
46. Lebl, M. (1999). *J. Comb. Chem.*, **1**, 3.
47. Kanie, O., Barresi, F., Ding, Y., Labbe, J., Otter, A., Forsberg, L. S., Ernst, B. and Hindsgaul, O. (1995). *Angew. Chem. Int. Ed. Engl.*, **34**, 2720.
48. Boons, G.-J., Heskamp, B. and Hout, F. (1996). *Angew. Chem. Int. Ed. Engl.*, **35**, 2845.
49. Liang, R., Yan, L., Loebach, J., Ge, M., Uozumi, Y., Sekanina, K., Horan, N., Gildersleeve, J., Thompson, C., Smith, A., Biswas, K., Still, W. C. and Kahne, D. (1996). *Science*, **274**, 1520.
50. Borchardt, A. and Still, W. C. (1994). *J. Am. Chem. Soc.*, **116**, 373.
51. Nestler, H. P., Bartlett, P. A. and Still, W. C. (1994). *J. Org. Chem.*, **59**, 4723.
52. Johnson, M., Arles, C. and Boons, G.-J. (1998). *Tetrahedron Lett.*, **39**, 9801.
53. Izumi, M. and Ichikawa, Y. (1998). *Tetrahedron Lett.*, **39**, 2079.
54. Wong, C.-H., Ye, X.-S. and Zhang, Z. (1998). *J. Am. Chem. Soc.*, **120**, 7137.
55. Zhu, T. and Boons, G.-J. (1998). *Angew. Chem. Int. Ed.*, **37**, 1898.

56. Wunberg, T., Kallus, C., Opatz, T., Henke, S., Schmidt, W. and Kunz, H. (1998). *Angew. Chem. Int. Ed.*, **37**, 2503.
57. Kobayashi, S. (1999). *Chem. Soc. Rev.*, **28**, 1.
58. St. Hilaire, P. M., Lowary, T. L., Meldal, M. and Bock, K. (1998). *J. Am. Chem. Soc.*, **120**, 13312.
59. Boons, G.-J. (1999). *Carbohydrates in Europe*, Dec. 27, p. 28.

Enzyme-catalysed Glycoside Synthesis[1-15]

There are very many enzymes that are associated with the processing of carbohydrates. From the very genesis of monosaccharides, ensconced in the process of photosynthesis, to their ultimate destruction, mainly through the glycolysis sequence, there are enzymes that hasten the processes of isomerization, oxidation, reduction, dehydration, phosphorylation and carbon–carbon bond formation. As well, other enzymes (synthases and transferases) are responsible for the polymerization of monosaccharides into polysaccharides or the synthesis of complex oligosaccharides; hydrolases, on the other hand, liberate the monosaccharide from a polysaccharide or "trim" the structure of one oligosaccharide into another. It is the transferases and hydrolases that will be discussed here because they are, together, so useful and important in the construction of the glycosidic linkage.

Glycosyl Transferases[16-25]

Glycosyl transferases are responsible for the transfer of a glycosyl unit from one oxygen atom to another (usually). In doing so, the donor is generally a glycoside, disaccharide, polysaccharide, glycosyl phosphate or glycosyl diphosphonucleoside;[26,27] in certain instances, the glycosyl unit may be attached, *via* phosphate or pyrophosphate diester linkages, to dolichol, undecaprenol or polyprenol carriers. The acceptor is usually the hydroxyl group of another sugar; when the acceptor is a water molecule, hydrolysis is the end result and the enzyme is better classed as a hydrolase.

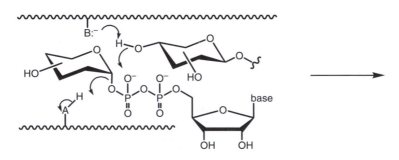

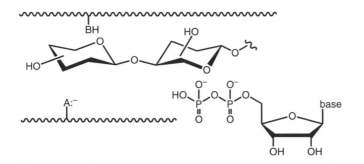

To date, there are three reported crystal structures for diphosphonucleoside-dependent glycosyl transferases[28-30] and the "mechanism" shown above, in fact, stems from knowledge gained of the mode of action of hydrolases (to be discussed next) and from indirect evidence; it is quite possible (probable) that the role of the general acid (AH) is often supplanted by a divalent metal ion. Nevertheless, glycosyl transferases are becoming increasingly important reagents for the synthesis of the glycosidic linkage in the laboratory. The main advantage of the method is the high regio- and stereo-selectivity exhibited by the enzymes, the disadvantages being the cost of the enzyme and the nucleoside donor and (sometimes) the unique substrate specificity exhibited by each enzyme.[31] Modern molecular biology has done much to relieve the problems by allowing access to gram quantities of the most important enzymes.[6,32,33] Two examples follow:

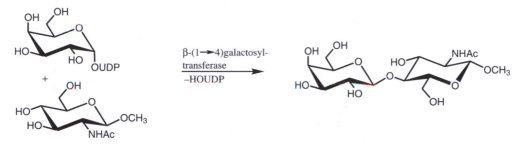

["product inhibition" by the liberated UDP (HOUDP) can be avoided by the use of a phosphatase (UDP → UMP + P$_i$)[34,35] or "cofactor regeneration" (UDP → UTP → UDPGlc → UDPGal)[6,36]]

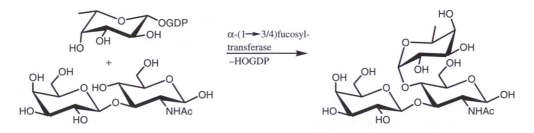

Glycoside Hydrolases (Glycosidases)[1,3,9,13-15,37-41]

Nature abounds with enzymes that can hydrolyse the glycosidic linkage. Virtually all organisms, from the lowly microbe to the omnipotent human, have the need to break down polysaccharides to give the small molecules that can cross membranes and so provide the energy to drive the biochemistry for survival. In addition, these same glycoside hydrolases, or glycosidases, are involved in a myriad of other processes that modify the biomolecules (glycoproteins, proteoglycans, glycolipids) necessary for a successful cell life. Glycosidases have been much more thoroughly studied than the less available glycosyl transferases —the knowledge gained has resulted in useful methods for the actual *synthesis* of the glycosidic linkage.

Classification of glycosidases: In a broad sense, glycosidases may be classified according to the substrate on which they act — we thus have amylases (amylose), cellulases (cellulose), xylanases (xylan), chitinases (chitin) and so on. More traditionally, and a little more accurately, glycosidases may be classified according to the type of bond hydrolysed; an α-D-galactosidase causes the cleavage of an α-D-galactopyranosidic linkage, the enzyme being stereospecific in its action in two senses (α and D). Following on from this came the IUBMB Enzyme Commission classification based on substrate specificity, with (EC 3.2.1.x) describing glycosidases.[42] The most recent (and parallel) system of classification is based on sequence similarities within the protein, allowing for the arrangement of nearly eighty different families of glycosidases;[43,44] glycosyl transferases/synthases have been similarly classified.[45,a] Not surprisingly, the members of each family have a similar protein "fold" and so present an active-site that ensures a similarity of mechanism (but not specificity) in the hydrolysis reaction. When several families with apparently unrelated sequence similarities show some uniformity in their three-dimensional structures (from X-ray crystallographic studies[46]), a "clan" is formed.[47,48]

Exo-, endo-, inverting and retaining glycosidases: The general reaction catalysed by a glycosidase is the hydrolysis of the anomeric linkage:

---

[a] http://afmb.cnrs-mrs.fr/~pedro/CAZY/db.html

On scrutiny of the products from the hydrolysis of a range of oligo- and polysaccharides, it appears that there exist two sorts of hydrolases: those that hydrolyse a glycosidic linkage at the end of a chain (*exo*-glycosidases), most commonly from the non-reducing end, and those that cleave an internal glycosidic linkage (*endo*-glycosidases):

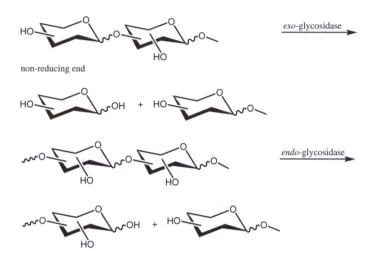

An even closer scrutiny of the products from such hydrolyses, using either *exo*- or *endo*-glycosidases, and of those products formed from simple glycosides, reveals that initially only one anomer of the free sugar is formed — so are described *inverting* and *retaining* glycosidases:

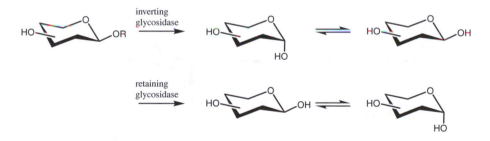

Certainly, the simplest way to determine whether a glycosidase is inverting or retaining is to employ nuclear magnetic resonance spectroscopy (Figure 5).[49,50] The enzyme in question, a glucanase, is an *endo*-hydrolase which clearly hydrolyses its substrate, a β-glucan (a β-linked polymer of D-glucose), with retention of configuration. Only a doublet at δ 4.67, characteristic of the β-anomer of the free sugar, appears after a few minutes and this is followed somewhat later by the appearance of another doublet (δ 5.23, the α-anomer of the same free sugar) from the process of mutarotation.

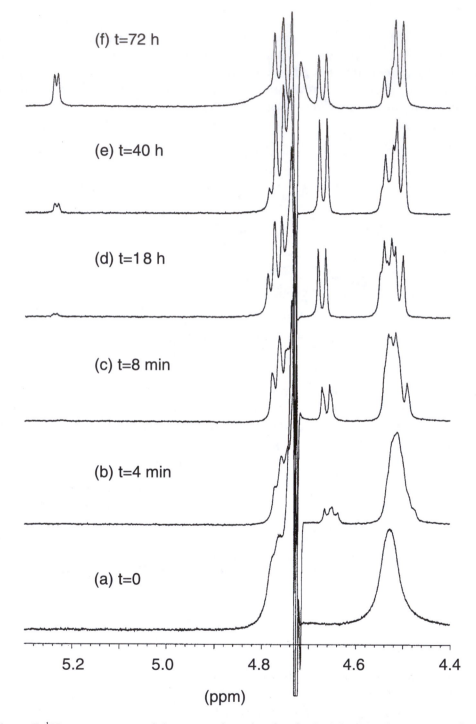

(f) t=72 h

(e) t=40 h

(d) t=18 h

(c) t=8 min

(b) t=4 min

(a) t=0

5.2   5.0   4.8   4.6   4.4

(ppm)

Figure 5 [1]H n.m.r. spectra of the anomeric region for the β–1,3–1,4–glucanase catalysed hydrolysis of barley β-glucan at different times. Reproduced with permission.[50]

With this distinction between inverting and retaining enzymes, it is now possible to understand (the somewhat archaic) terms such as "α-amylase" and "β-amylase"; the former describes a hydrolase that acts on amylose (an α-linked polymer of D-glucose) to produce free sugars with retention of configuration, whereas the latter conducts a similar hydrolysis but with inversion of configuration. Although not mentioned earlier, glycosyl transferases and glycoside synthases may retain the stereochemistry of the relevant donor in the product or, conversely, exhibit an inversion.

The mechanism of action of glycosidases:[51,52] How can we rationalize the stereoselectivity (inverting or retaining) exhibited by glycosidases? Koshland made the first sensible proposals for the mechanism of action of glycosidases in 1953 and, really, not much has changed since then:[53]

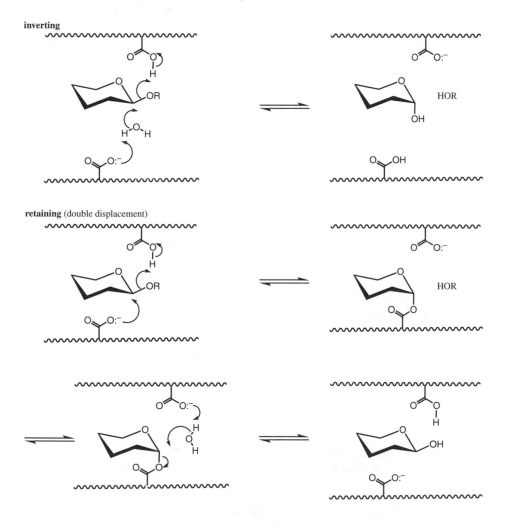

Not surprisingly, an enormous amount of effort has gone into establishing the two proposals above; X-ray crystallography has shown the different dimensions of the active sites of inverting and retaining glycosidases (9.5 Å and 5.5 Å, respectively),[54,55] various studies have shown that adjacent residues in the enzyme are able to modulate the pK_a of the general acid/base residue[56] and experiments with special substrates, so designed as to be active in the initial step and then much less active in the subsequent hydrolysis step, have been successful in routinely labelling the "catalytic nucleophile" of many retaining glycosidases:[57-59]

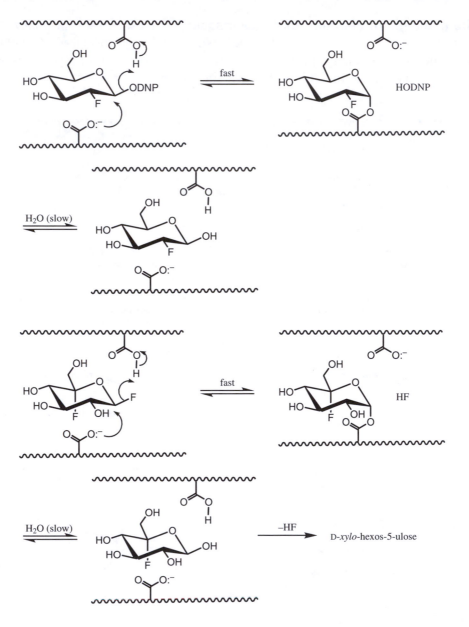

A new concept introduced by Vasella, that of in-plane (*syn* or *anti*) protonation of the substrate by the carboxylic acid at the catalytic site of the enzyme, has added another dimension to our understanding of the mechanism of action of glycoside hydrolases.[60,61]

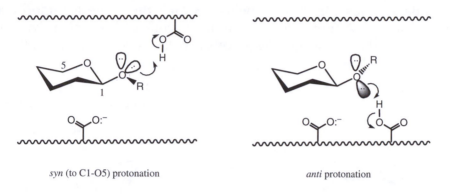

| *syn* (to C1-O5) protonation | *anti* protonation |

We have now discussed enough about the nature of glycosidases to understand their employment in the synthesis of the glycosidic linkage. Fortunately, these enzymes are ubiquitous (and hence readily available), cheap and robust and show absolute stereoselectivity in their construction of the glycosidic linkage; on the other hand, they sometimes exhibit poor regioselectivity and, partly as a consequence of this trait, the chemical yield of glycoside can be quite low.

Glycoside synthesis by "reverse(d) hydrolysis" (equilibrium-controlled synthesis, the "thermodynamic" approach): The hydrolysis of a glycoside, although not a spontaneous process *in vivo*, is generally regarded as an irreversible process that gives good yields of products:

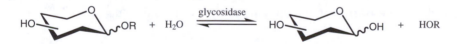

Therefore, if one wishes to synthesize glycosides by reversing this process, various conditions must be employed to control the equilibrium in the desired direction. Moderate success has been achieved by increasing the concentration of both the alcohol acceptor and the sugar donor, by reducing the amount and activity of the water (adding an organic co-solvent) and varying the temperature (the hydrolysis reaction is an exothermic process):[13]

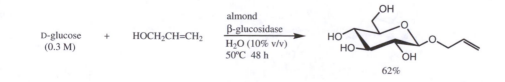

The main advantage of the process is its simplicity; generally, the yields are somewhat low (15–40%) and separation of the product from the excess of starting materials is usually tiresome.

Glycoside synthesis by "transglycosylation" (the "kinetic" approach):
If one reconsiders the proposed mechanism of action of a retaining glycosidase, the reversal of the first step of the process should provide a viable synthesis of the glycosidic linkage:

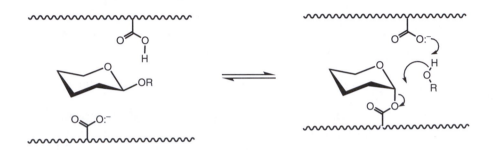

For such a synthesis to be successful, the "glycosyl enzyme" intermediate must be formed efficiently from some reactive donor, the glycosyl residue must be transferred rapidly to an acceptor alcohol (rather than to water) and, in deference to this point, there must be a binding site in the enzyme to accommodate the acceptor alcohol.

Generally, the method involves the treatment of an excess of a glycosyl fluoride, 4-nitrophenyl glycoside or disaccharide (of the same anomeric configuration as that desired in the product) with the acceptor alcohol (often a glycoside itself) in a mixture of water and organic solvent containing the glycosidase (the purity of the enzyme is not a critical issue here):[62,63]

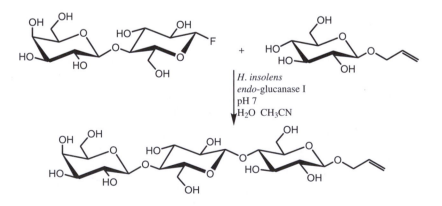

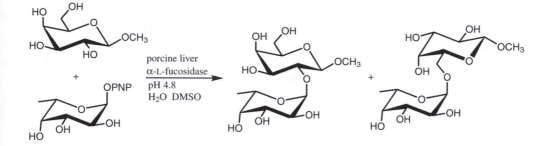

The yields of products tend to be higher (40–50%) than in the "reverse hydrolysis" approach but, because of the infidelity forced on the enzyme (the normal acceptor is water), there is often the complication of the formation of regioisomers. This same situation can arise when the same enzyme from different sources is used. A "multiple enzyme" approach can aid in the purification step:[64]

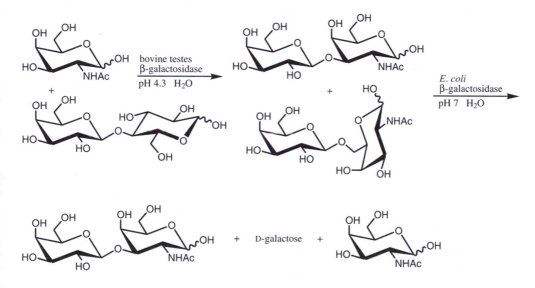

The bottom line on this general method of glycoside synthesis is that one must "strike while the iron is hot", *viz.* each reaction mixture must be closely monitored so that the desired product can be harvested at a particular time before deleterious rearrangements, hydrolysis and over-glycosidation can occur. Some recent examples may prove interesting.[65–71]

Glycoside synthesis using "glycosynthases" (mutant glycosidases):

This last approach to the synthesis of glycosides utilizing enzyme assistance is the brainchild of Steve Withers — by his own account, while cycling home one evening from the University of British Columbia, he realized that a glycosidase so-engineered that it lacked the catalytic nucleophile (COOH) could be given a glycosyl fluoride of the *wrong* configuration and made to believe that it had formed the normal glycosyl enzyme intermediate. In the presence of an acceptor alcohol, the remaining machinery of the enzyme would form a new glycoside!

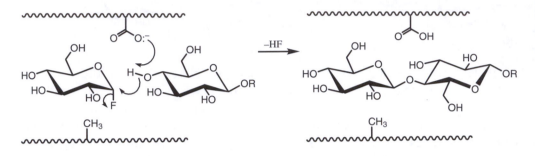

The beauty of this method is that the mutant enzyme, the "glycosynthase", lacks the machinery to catalyse the hydrolysis of any glycosidic linkage present in the product.

In the original concept of the method,[72] an *exo*-retaining β-glucosidase (Abg) from an *Agrobacterium* sp. was subjected to site-directed mutagenesis to convert the catalytic nucleophile, glutamate 358 into alanine (Glu358Ala). When this mutant was incubated with various acceptors (aryl glycosides were found to be the best), good to excellent yields of glycosides resulted:

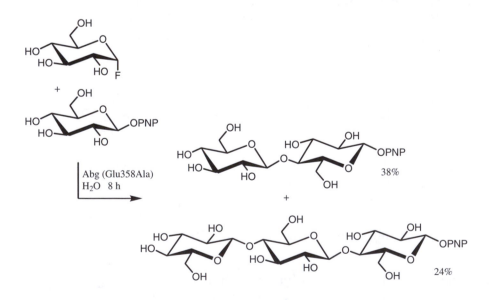

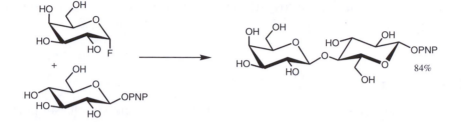

Recent work has focused on other mutants of Abg and the serine mutant (Glu358Ser) has exhibited similar glycosynthase activity but with much shortened reaction times (a matter of a few hours).[73] Also, a mutant mannosidase is under investigation.[74]

Planas has independently developed a mutant (Glu134Ala) of the *endo*-retaining 1,3-1,4-β-glucanase from *Bacillus licheniformis* and shown its use in the synthesis of oligosaccharides:[75]

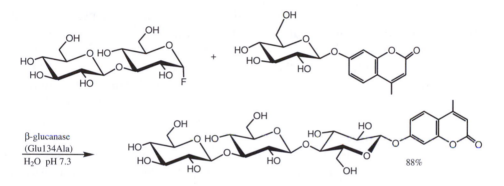

Driguez has also obtained a mutant (Glu197Ala) of the *endo*-retaining β-glucosidase (EGI) from *Humicola insolens* and it has been used successfully in the glycosylation of a range of acceptor alcohols:[76]

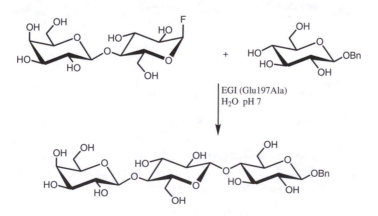

There is no doubt that glycosynthases will play a major role in the synthesis of oligosaccharides in the future, both commercially and for research purposes.[77]

Other Matters

The attachment of the substrate or the enzyme to a polymeric support has greatly assisted the enzyme-catalysed synthesis of glycosides. On the one hand, a series of enzymes in solution may be used to add monosaccharide units to a glyco-polymer primer, itself either insoluble beads or soluble — cleavage from the polymer then releases the synthetic oligosaccharide.[13,78] Alternatively, the enzyme itself may be immobilized on a polymer (agarose, polyacrylamide, glass) or located in a dialysis bag and so, after the required glycoside synthesis, easily recovered.[2,79,80] Tabular surveys of the commercially available enzymes and glycosidase catalyzed syntheses (now numbering over 1000) have recently appeared.[39,81]

Another intense area of research, both in synthetic and biochemical terms, involves the inhibition of carbohydrate-processing enzymes, particularly the glycosidases and, more of late, the glycosyl transferases and glycoside synthases. Such research has application in the understanding of the mechanism of action of many enzymes used in industry (pulp bleaching, biomass degradation, digestibility of foods) and the development of drugs to combat human diseases (diabetes, AIDS, cancer, influenza). Apart from the 2-deoxy-2-fluoro- and 5-fluoroglycosyl derivatives which were mentioned previously and are capable of both labelling and inhibiting a wide range of retaining glycosidases, there exists a totally different class of glycosidase inhibitor which is based on polyhydroxylated, nitrogen-containing heterocycles:

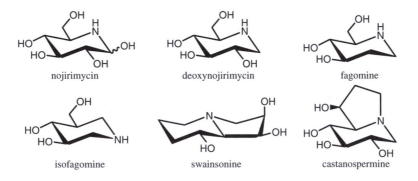

nojirimycin   deoxynojirimycin   fagomine

isofagomine   swainsonine   castanospermine

Much of the ability of these heterocycles to act as competitive inhibitors of glycosidases is attributed to their shape, basicity and tight binding of the conjugate acid to the active-site of the enzyme.[60,61,82–91]

References

1. Drueckhammer, D. G., Hennen, W. J., Pederson, R. L., Barbas, C. F., III, Gautheron, C. M., Krach, T. and Wong, C.-H. (1991). *Synthesis*, 499.
2. David, S., Augé, C. and Gautheron, C. (1991). *Adv. Carbohydr. Chem. Biochem.*, **49**, 175.
3. Bednarski, M. D. and Simon, E. S. eds (1991). *Enzymes in Carbohydrate Synthesis*, ACS Symposium Series 466. American Chemical Society, Washington, DC.
4. Wong, C. H. and Whitesides, G. M. (1994). *Enzymes in Synthetic Organic Chemistry*, Tetrahedron organic chemistry series, Baldwin, J. E. and Magnus, P. D. eds, vol. 12, Pergamon, Oxford, p. 252.
5. Drauz, K. and Waldmann, H. eds (1995). *Enzyme Catalysis in Organic Synthesis. A Comprehensive Handbook*, vol. I, VCH, Weinheim, p. 165.
6. Wong, C.-H., Halcomb, R. L., Ichikawa, Y. and Kajimoto, T. (1995). *Angew. Chem. Int. Ed. Engl.*, **34**, 412, 521.
7. Waldmann, H. (1995). "Enzymatic synthesis of *O*-glycosides", in *Organic Synthesis Highlights II*, Waldmann, H. ed., VCH, Weinheim, p. 157.
8. Copeland, R. A. (1996). *Enzymes. A Practical Introduction to Structure, Mechanism, and Data Analysis*, VCH, Weinheim.
9. Křen, V. and Thiem, J. (1997). *Chem. Soc. Rev.*, **26**, 463.
10. Takayama, S., McGarvey, G. J. and Wong, C.-H. (1997). *Chem. Soc. Rev.*, **26**, 407.
11. Fitz, W. and Wong, C.-H. (1997). Oligosaccharide synthesis by enzymatic glycosidation, in *Preparative Carbohydrate Chemistry*, Hanessian, S. ed., Marcel Dekker, New York, p. 485.
12. Augé, C. and Crout, D. H. G. eds (1997). *Carbohydr. Res.*, **305**, 307–568.
13. Crout, D. H. G. and Vic, G. (1998). *Curr. Opin. Chem. Biol.*, **2**, 98.
14. Davies, G., Sinnott, M. L. and Withers, S. G. (1998). Glycosyl transfer, in *Comprehensive Biological Catalysis. A Mechanistic Reference*, Sinnott, M. ed., vol. 1, Academic Press, London, p. 119.
15. Zechel, D. L. and Withers, S. G. (1999). Glycosyl transferase mechanisms, in *Comprehensive Natural Products Chemistry*, Sir Derek Barton, Nakanishi, K., Meth-Cohn, O. and Poulter, C. D. eds, vol. 5, Elsevier, Amsterdam, p. 279.
16. Watkins, W. M. (1986). *Carbohydr. Res.*, **149**, 1.
17. Paulson, J. C. and Colley, K. J. (1989). *J. Biol. Chem.*, **264**, 17615.
18. Crawley, S. C. and Palcic, M. M. (1996). Use of glycosyltransferases in the synthesis of unnatural oligosaccharide analogues, in *Modern Methods in Carbohydrate Synthesis*, Khan, S. H. and O'Neill, R. A. eds, Harwood Academic, Netherlands, p. 492.
19. Wong, C.-H. (1996). Practical synthesis of oligosaccharides based on glycosyltransferases and glycosylphosphites, in *Modern Methods in Carbohydrate Synthesis*, Khan, S. H. and O'Neill, R. A. eds, Harwood Academic, Netherlands, p. 467.
20. Matta, K. L. (1996). Synthetic glycosyltransferase acceptors and inhibitors; useful tools in glycobiology, in *Modern Methods in Carbohydrate Synthesis*, Khan, S. H. and O'Neill, R. A. eds, Harwood Academic, Netherlands, p. 437.
21. Guo, Z. and Wang, P. G. (1997). *Appl. Biochem. Biotechnol.*, **68**, 1.
22. Gambert, U. and Thiem, J. (1997). *Topics Curr. Chem.*, **186**, 21.
23. Öhrlein, R. (1999). *Topics Curr. Chem.*, **200**, 227.
24. Gambert, U. and Thiem, J. (1999). *Eur. J. Org. Chem.*, 107.
25. Stone, B. A. and Svensson, B. Enzymic depolymerisation and synthesis of polysaccharides, in *Glycoscience – Chemistry and Chemical Biology*, Thiem, J., Fraser-Reid, B. O. and Tatsuta, K. T. eds. vol. 3, Springer-Verlag.
26. Caputto, R., Leloir, L. F., Cardini, C. E. and Paladini, A. C. (1950). *J. Biol. Chem.*, **184**, 333.

27. Leloir, L. F. (1971). *Science*, **172**, 1299.
28. Vrielink, A., Rüger, W., Driessen, H. P. C. and Freemont, P. S. (1994). *EMBO J.*, **13**, 3413.
29. Charnock, S. J. and Davies, G. J. (1999). *Biochemistry*, **38**, 6380.
30. Gastinel, L. N., Cambillau, C. and Bourne, Y. (1999). *EMBO J.*, **18**, 3546.
31. Kajihara, Y., Kodama, H., Endo, T. and Hashimoto, H. (1998). *Carbohydr. Res.*, **306**, 361.
32. Tsuji, S. (1996). *J. Biochem.*, **120**, 1.
33. Drauz, K. and Waldman, H. eds (1995). *Enzyme Catalysis in Organic Synthesis: A Comprehensive Handbook*, vol. II, VCH, Weinheim, p. 929.
34. Unverzagt, C., Kunz, H. and Paulson, J. C. (1990). *J. Am. Chem. Soc.*, **112**, 9308.
35. Unverzagt, C., Kelm, S. and Paulson, J. C. (1994). *Carbohydr. Res.*, **251**, 285.
36. Wong, C.-H., Haynie, S. L. and Whitesides, G. M. (1982). *J. Org. Chem.*, **47**, 5416.
37. Withers, S. G. (1999). *Can. J. Chem.*, **77**, 1.
38. Fernández-Mayoralas, A. (1997). *Topics Curr. Chem.*, **186**, 1.
39. Vocadlo, D. J. and Withers, S. G. (2000). Glycosidase-catalyzed oligosaccharide synthesis, in *Carbohydrates in Chemistry and Biology*, Ernst, B., Hart, G. W. and Sinaÿ, P. eds, vol. 2, Wiley-VCH, p. 723.
40. Ly, H. D. and Withers, S. G. (1999). *Annu. Rev. Biochem.*, **68**, 487.
41. Nilsson, K. G. I. (1996). Synthesis with glycosidases, in *Modern Methods in Carbohydrate Synthesis*, Khan, S. H. and O'Neill, R. A. eds, Harwood Academic, Netherlands, p. 518.
42. Webb, E. C. ed. (1992). *Enzyme Nomenclature 1992. Recommendations of the Nomenclature Committee of the International Union of Biochemistry and Molecular Biology on the Nomenclature and Classification of Enzymes*, Academic Press, London.
43. Henrissat, B. and Davies, G. (1997). *Curr. Opin. Struct. Biol.*, **7**, 637.
44. Claeyssens, M. and Henrissat, B. (1992). *Protein Sci.*, **1**, 1293.
45. Campbell, J. A., Davies, G. J., Bulone, V. and Henrissat, B. (1997). *Biochem. J.*, **326**, 929.
46. Davies, G. and Henrissat, B. (1995). *Structure*, **3**, 853.
47. Henrissat, B., Callebaut, I., Fabrega, S., Lehn, P., Mornon, J.-P. and Davies, G. (1995). *Proc. Natl. Acad. Sci. USA*, **92**, 7090.
48. Jenkins, J., Leggio, L. L., Harris, G. and Pickersgill, R. (1995). *FEBS Lett.*, **362**, 281.
49. Withers, S. G., Dombroski, D., Berven, L. A., Kilburn, D. G., Miller, R. C. Jr Warren, R. A. J. and Gilkes, N. R. (1986). *Biochem. Biophys. Res. Commun.*, **139**, 487.
50. Malet, C., Jimínez-Barbero, J., Bernabé, M., Brosa, C. and Planas, A. (1993). *Biochem. J.*, **296**, 753.
51. Sinnott, M. L. (1990). *Chem. Rev.*, **90**, 1171.
52. Zechel, D. L. and Withers, S. G. (2000). *Acc. Chem. Res.*, **33**, 11.
53. Koshland, D. E. Jr (1953). *Biol. Rev.*, **28**, 416.
54. Wang, Q., Graham, R. W., Trimbur, D., Warren, R. A. J. and Withers, S. G. (1994). *J. Am. Chem. Soc.*, **116**, 11594.
55. Notenboom, V., Birsan, C., Warren, R. A. J., Withers, S. G. and Rose, D. R. (1998). *Biochemistry*, **37**, 4751.
56. McIntosh, L. P., Hand, G., Johnson, P. E., Joshi, M. D., Körner, M., Plesniak, L. A., Ziser, L., Wakarchuk, W. W. and Withers, S. G. (1996). *Biochemistry*, **35**, 9958.
57. Withers, S. G. and Aebersold, R. (1995). *Protein Sci.*, **4**, 361.
58. Davies, G. J., Mackenzie, L., Varrot, A., Dauter, M., Brzozowski, A. M., Schülein, M. and Withers, S. G. (1998). *Biochemistry*, **37**, 11707.
59. Sulzenbacher, G., Mackenzie, L. F., Wilson, K. S., Withers, S. G., Dupont, C. and Davies, G. J. (1999). *Biochemistry*, **38**, 4826.

60. Heightman, T. D. and Vasella, A. T. (1999). *Angew. Chem. Int. Ed.*, **38**, 750.
61. Varrot, A., Schülein, M., Pipelier, M., Vasella, A. and Davies, G. J. (1999). *J. Am. Chem. Soc.*, **121**, 2621.
62. Fairweather, J. K., Stick, R. V., Tilbrook, D. M. G. and Driguez, H. (1999). *Tetrahedron*, **55**, 3695.
63. Svensson, S. C. T. and Thiem, J. (1990). *Carbohydr. Res.*, **200**, 391.
64. Hedbys, L., Johansson, E., Mosbach, K., Larsson, P.-O., Gunnarsson, A. and Svensson, S. (1989). *Carbohydr. Res.*, **186**, 217.
65. Shoda, S., Fujita, M. and Kobayashi, S. (1998). *Trends Glycosci. Glycotech.*, **10**, 279.
66. Li, J., Robertson, D. E., Short, J. M. and Wang, P. G. (1998). *Tetrahedron Lett.*, **39**, 8963.
67. Hrmova, M., Fincher, G. B., Viladot, J.-L., Planas, A. and Driguez, H. (1998). *J. Chem. Soc., Perkin Trans. 1*, 3571.
68. Viladot, J.-L., Stone, B., Driguez, H. and Planas, A. (1998). *Carbohydr. Res.*, **311**, 95.
69. Fang, J., Xie, W., Li, J. and Wang, P. G. (1998). *Tetrahedron Lett.*, **39**, 919.
70. Endo, T., Koizumi, S., Tabata, K., Kakita, S. and Ozaki, A. (1999). *Carbohydr. Res.*, **316**, 179.
71. Scigelova, M., Singh, S. and Crout, D. H. G. (1999). *J. Chem. Soc., Perkin Trans. 1*, 777.
72. Mackenzie, L. F., Wang, Q., Warren, R. A. J. and Withers, S. G. (1998). *J. Am. Chem. Soc.*, **120**, 5583; US Patent 5 952 203.
73. Mayer, Ch., Zechel, D. L., Reid, S., Warren, R. A. J. and Withers, S. G. (2000). *FEBS Lett.*, **466**, 40.
74. Nashiru, O., Zechel, D.L., Stoll, D., Mohammadzadeh, T., Warren, R.A.J. and Withers, S.G. *Angew. Chem. Int. Ed.*, in press.
75. Malet, C. and Planas, A. (1998). *FEBS Lett.*, **440**, 208.
76. Fort, S., Boyer, V., Greffe, L., Davies, G. J., Moroz, O., Christiansen, L., Schülein, M., Cottaz, S. and Driguez, H. (2000). *J. Am. Chem. Soc.*, **122**, 5429.
77. Trincone, A., Perugino, G., Rossi, M. and Moracci, M. (2000). *Bioorg. Med. Chem. Lett.*, **10**, 365.
78. Osborn, H. M. I. and Khan, T. H. (1999). *Tetrahedron*, **55**, 1807.
79. Tischer, W. and Wedekind, F. (1999). *Topics Curr. Chem.*, **200**, 95.
80. Drauz, K. and Waldmann, H. eds (1995). *Enzyme Catalysis in Organic Synthesis. A Comprehensive Handbook*, vol. I, VCH, Weinheim, p. 73.
81. Drauz, K. and Waldmann, H. eds (1995). *Enzyme Catalysis in Organic Synthesis. A Comprehensive Handbook*, vol. II, VCH, Weinheim, p. 963.
82. Legler, G. (1990). *Adv. Carbohydr. Chem. Biochem.*, **48**, 319.
83. Jacob, G. S. (1995). *Curr. Opin. Struct. Biol.*, **5**, 605.
84. Bols, M. (1998). *Acc. Chem. Res.*, **31**, 1.
85. Fleet, G. W. J., Estevez, J. C., Smith, M. D., Blériot, Y., de la Fuente, C., Krülle, T. M., Besra, G. S., Brennan, P. J., Nash, R. J., Johnson, L. N., Oikonomakos, N. G. and Stalmans, W. (1998). *Pure Appl. Chem.*, **70**, 279.
86. Stütz, A. E. ed. (1999). *Iminosugars as Glycosidase Inhibitors. Nojirimycin and Beyond*, Wiley-VCH, Weinheim.
87. Ichikawa, Y., Igarashi, Y., Ichikawa, M. and Suhara, Y. (1998). *J. Am. Chem. Soc.*, **120**, 3007.
88. Kaneko, M., Kanie, O., Kajimoto, T. and Wong, C.-H. (1997). *Bioorg. Med. Chem. Lett.*, **7**, 2809.
89. Sears, P. and Wong, C.-H. (1998). *Chem. Commun.*, 1161.
90. Wong, C.-H. (1999). *Acc. Chem. Res.*, **32**, 376.

91. Asano, N., Nash, R. J., Molyneux, R. J. and Fleet, G. W. J. (2000). *Tetrahedron: Asymmetry*, **11**, 1645.

Chapter 10

Disaccharides, Oligosaccharides and Polysaccharides[1]

A book such as this now finds itself "between the Devil and the deep blue sea" — there must be mention of disaccharides, oligosaccharides and polysaccharides, but at what level? There are annual meetings of learned societies where the whole discussion, for up to a week, centres around just one carbohydrate, for example, lactose, sucrose, starch, cellulose or chitin!

What follows will be a listing of the more important di-, oligo- and polysaccharides, together with some pertinent chemical comments and, where possible, recent literature reviews.

Disaccharides

Reducing Disaccharides

Maltose is produced by the enzymic hydrolysis (amylases) of starch and is a precursor of D-glucose in the mammalian digestive tract:

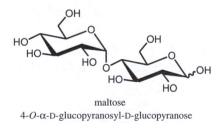

maltose
4-*O*-α-D-glucopyranosyl-D-glucopyranose

"Maltose syrup" is produced by the enzymic hydrolysis of starch and contains about 60% maltose, 25% D-glucose and 15% of maltodextrins; it is used as a sweetener, mainly in foods, boosting "mouth feel", and as a fermentation substrate in brewing.

Isomaltose is a regioisomer of maltose and constitutes the branch point of the polysaccharides, amylopectin and glycogen:

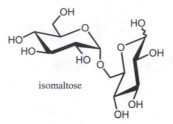

isomaltose

Cellobiose is a product of the bacterial hydrolysis of cellulose, these organisms possessing the necessary enzymes (cellobiohydrolases and *endo*-cellulases) to carry out the degradation:

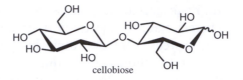

cellobiose

The octaacetate of cellobiose is easily obtained by the treatment of cotton or filter paper with acetic anhydride and sulfuric acid.

Gentiobiose is the carbohydrate portion of a large number of glycosides, one of which (amygdalin) we have already encountered:

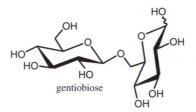

gentiobiose

Laminaribiose is the repeating unit found in the polysaccharides, laminarin (brown algae), pachyman (fungi), paramylon (unicellular algae) and callose:

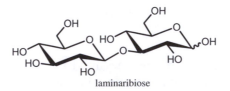

laminaribiose

Xylobiose is the repeating unit of various polysaccharides, for example, the xylans found in plant cell walls:

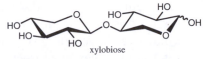

xylobiose

Mannobiose is the repeating unit of the mannans, again plant polysaccharides:

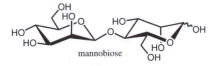

mannobiose

Lactose, a by-product in the manufacture of cheese, is the main disaccharide found in the milk of mammals (cow, 4.5% w/v; human, 6.5% w/v) and contains both D-galactose and D-glucose residues:

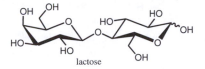

lactose

Some ethnic groups lose the enzyme "lactase" after infancy and become "lactose intolerant"; others that cannot process the D-galactose (to D-glucose) released from lactose on hydrolysis develop the symptoms of "galactosaemia". Both medical conditions can be remedied by a lactose-free diet.

Non-reducing Disaccharides

There is a genuine difference in function between reducing and non-reducing disaccharides. The former, containing a reactive hemiacetal centre, can easily be exploited chemically (oxidation, reduction, imine formation with an amine); the latter, however, are usually stored by plants and can be translocated in the plant.

Sucrose is a non-reducing disaccharide that is the main sugar in the extract of sugar cane and sugar beet:

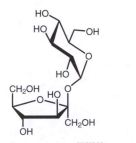

sucrose

β-D-fructofuranosyl α-D-glucopyranoside

Sucrose is still the world's main sweetening agent and some 10^8 tons are produced annually. The plant enzyme, invertase, is able to hydrolyse sucrose ($[\alpha]_D$ +66°) into its two constituent sugars and the equimolar mixture of D-fructose and D-glucose so produced is termed "invert sugar" ($[\alpha]_D$ −22°).

Trehalose, a molecule that we have already encountered, is another constituent of this class and is found in various insects, fungi and microbes:

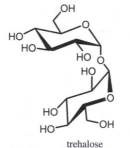

trehalose
α-D-glucopyranosyl α-D-glucopyranoside

Oligosaccharides[2]

Raffinose is a non-reducing trisaccharide found in plants and is a D-galactosylated version of sucrose: α-D-galactopyranosyl-(1→6)-α-D-glucopyranosyl β-D-fructofuranoside, abbreviated as α-D-Gal*p*-(1→6)-α-D-Glc*p*-(1↔2)-β-D-Fru*f*. Many other oligosaccharides, however, are not found as discrete molecules but are joined (glycoconjugates) to other biomolecules (proteins and lipids, in particular). The archetypal examples are the *blood-group substances*, those oligosaccharides that are found conjugated to a lipid (glycolipid) on the surface of the erythrocyte [and to a protein (glycoprotein) on the surface of other cells] and are responsible for the A, B, AB and O blood group serotypes:

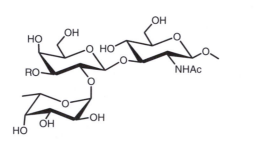

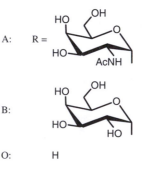

Type A blood contains antibodies to type B and *vice versa* so that cross-transfusion is not possible. Type AB blood contains both the A and B antigenic determinants and so such carriers are universal acceptors; type O blood does not elicit an immune response in type A, B or AB carriers and so such people are universal donors.

The *cyclodextrins*, α-, β- and γ-, are a group of cyclic oligosaccharides that contain six, seven and eight sugar rings, respectively, and are formed by the enzymic degradation of starch. As such, these molecules consist of a ring of six to eight D-glucopyranose units held together by α-1,4-linkages; the result is a bottomless "tub" with a hydrophilic exterior and a hydrophobic interior, into which "guest" molecules can fit. There has been an enormous amount of work done on these cyclodextrin "host" molecules and they are used in both the academic laboratory and in industrial processes.[3]

Polysaccharides[4–9]

The polysaccharides are an amazing group of biopolymers that have evolved to serve a whole range of functions in the host organism — energy storage, structure, defence, pathogenicity and wound healing, to name but a few. Very often, a subtle change in the structure of the monomer or the type of glycosidic linkage employed has a profound effect on the properties and function of the resultant polysaccharide.

Starch[10,11]

Starch is produced by all higher plants and is stored as granules in the roots, tubers, rhizomes, fruit and seeds as an energy reserve. There are reports on starch dating back thousands of years, mainly concerned with its use as a glue and sizing agent in the early production of paper. Starch did not enter the world of commerce until much later, being well established in about the fourteenth century. Most starch is now produced from maize but other sources such as potato and, of late, banana give a product with individual characteristics that are suited to various aspects of the food and related industries — you may remember that "levoglucosan" (1,6-anhydro-β-D-glucopyranose) is best produced by the pyrolysis of potato starch.

Starch actually consists of two polymers, amylose (20%) and amylopectin (80%). Some specially bred or engineered forms of plants can produce high-amylose starch (low amylopectin content) or amylose-free starch (contains mainly amylopectin), an example of the latter being glutinous rice which yields "waxy starch". Again, these starches have specialty uses in the food and related industries.

Amylose is a relatively low molecular weight ($\sim 10^6$) polymer which has maltose as its "repeating unit":

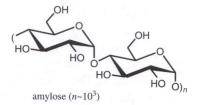

amylose ($n \sim 10^3$)

Amylose, as a result of the inherent $(1 \rightarrow 4)$-α-D linkage, takes on a helical or coiled tube shape and is semi-crystalline. The helix has ideal dimensions to accept an iodine guest molecule and this interaction is responsible for the familiar blue colouration.[12]

Amylopectin (molecular weight $\sim 10^8$) is a much larger polymer than amylose, having the same repeating unit as the latter but with branching $[(1 \rightarrow 6)$-α-$]$ about every ten maltose units; amylopectin thus contains the elements of both maltose and isomaltose:

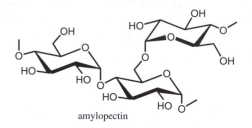

amylopectin

Not surprisingly, amylopectin is globular in shape and relatively non-crystalline.

Glycogen

Glycogen in mammals is the counterpart of starch in plants — it is a storage "powerhouse" of energy. The structure of glycogen is much akin to that of amylopectin except that the branching is more frequent, about every five maltose units, and the molecular weight may be even higher. Various debranching enzymes and glycogen phosphorylase combine in times of stress or need to produce D-glucose 1-phosphate for entry into the glycolysis sequence.

Cellulose[13–15]

Cellulose is the most abundant polymer on Earth and constitutes such an important part of the biomass that an enormous effort has been put into understanding its biosynthesis, function and degradation. Cellulose is a polymer, of varying molecular weight, of cellobiose units:

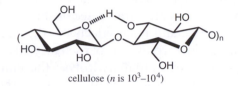

cellulose (n is 10^3–10^4)

As such, with the rotation exhibited within each cellobiose unit (stabilized by hydrogen bonding), the cellulose is able to take on a "twisted ribbon" structure. Hydrogen bonding between chains, in either a parallel or anti-parallel sense (giving two types of cellulose, namely cellulose I and cellulose II), imparts more rigidity to the polymer and a subsequent packaging of these bound chains into "microfibrils" and then fibres forms the ultimate in building material for Nature – cellulose is the major structural material of the cell wall of plants and a host of other organisms, for example, brown algae and some fungi. Cotton fibres (seed hairs) are nearly pure cellulose.

One of the main uses of cellulose, of course, is in the pulp and paper industry where there is great interest in replacing the traditional bleaching processes with enzyme-based procedures. There is much effort being expended in understanding and harnessing the workings of the "cellulosome", that collection of proteins and enzymes (cellulases) found on the surface of certain bacteria that is able to degrade cellulose in its natural environment.[16] Issues here are also critical to recycling programmes and the enormous textile industry based on cotton.

It is a curiosity of evolution that humans utilize starch as a primary source of energy, having the ability to hydrolyse the α-D-glucosidic linkage, but are unable to utilize cellulose in anything but a second-order nutritional role. Also, some interesting and recent comments have been made on the biosynthesis of cellulose.[17]

Chitin[18–21]

If cellulose can be described as the bulwark of the plant kingdom, then surely chitin must be the counterpart in the animal world. Found mainly in the shells of crustaceans, but also in the exoskeleton of insects and the cell wall of many fungi (20% in mushrooms), chitin is a polymer of N-acetyl-D-glucosamine with N^I,N^{II}-diacetylchitobiose being the repeat unit:

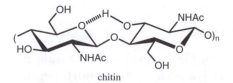

chitin

The structure of chitin is somewhat reminiscent of that of cellulose, with the "rotation" and intramolecular hydrogen bond quite obvious; the molecular weight of chitin also approximates that of cellulose. The presence of the acetamido group in chitin, however, confers an extra degree of stability to the polymer owing to a combination of electronic and polar effects — the intermolecular hydrogen bonds involving the acetamido groups are particularly strong.

Enzymes (chitinases) do exist that hydrolyse the chitin chain (some into N^I,N^{II}-diacetylchitobiose) in natural events such as the defence by plants against fungal pathogen attack. Kobayashi, using a chitinase in the transglycosylation mode, has reported a spectacular synthesis of chitin from an activated chitobiose derivative:[22]

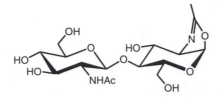

Heparin[23-27]

Heparin is a heteropolysaccharide that prevents clots from forming in blood vessels. It is found principally in cells that line arterial walls and is used intravenously after surgery and for the treatment of thrombosis.

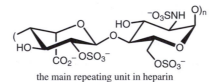

the main repeating unit in heparin

Heparin is a highly sulfated and hence highly charged polymer that brings antithrombin, a key inhibitor of the blood coagulation cascade, into close proximity to the clot-forming protein, thrombin.[28] Heparin is expensive and not without side effects (bleeding, platelet reduction), encouraging an enormous search for a simpler and more satisfactory drug. Recently, an oligosaccharide has been synthesized which retains the two binding domains of heparin, is more specific in its action and is ten times more potent than heparin.[29]

Bacterial Polysaccharides[30,31]

The cell envelope of both Gram-positive and Gram-negative bacteria is based on peptidoglycan, a polymer in which short peptide chains cross-link the polysaccharide chains to each other. Additionally, in Gram-negative bacteria, there is an outer membrane that is composed of a lipopolysaccharide–protein complex. In Gram-positive bacteria, there is no outer membrane but the peptidoglycan wall is thicker and has attached to it a polysaccharide and teichoic acid (a diester polymer of phosphoric acid and various polyols and sugars). Both sorts of bacteria produce extracellular polysaccharides, present as either a discrete capsule covalently attached to the cell envelope or as a slime weakly bound to the cell surface. These various glycoconjugates and polysaccharides on the surface of the cell constitute the antigenic determinants that initiate an immunogenic response in a host; in addition, these same surface carbohydrates provide the recognition element for pathogens of the bacterium, such as bacteriophages.

The lipopolysaccharide[32] of Gram-negative bacteria contains both a lipid (lipid A,[33] bound to the surface of the bacterium) and a polysaccharide (connected to the lipid by an inner and outer core) that are together the immunogens responsible for the O-specific serotypes. The capsular polysaccharides of both types of bacterium also elicit an immune response, the so-called K-antigens.

The structures of the various O-specific polysaccharides and K-antigens are unique, often being characterized by repeating "blocks" in the polymer structure.[34] Indeed, all types of monosaccharides, often including L-rhamnose (6-deoxy-L-mannose) and L-fucose (6-deoxy-L-galactose), are found, together with rarer, modified sugars such as 3,6-dideoxy hexoses and "Kdo" (3-deoxy-D-*manno*-octulosonic acid).[35,36]

We will return to this subject of antigenicity and the related immune response when we discuss the concept of carbohydrate-based vaccines in the final chapter.

References

1. Pigman, W. and Horton, D. eds (1970). *The Carbohydrates. Chemistry and Biochemistry*, vols. IIA, IIB, Academic Press, London.
2. Staněk, J., Černý, M. and Pacák, J. (1965). *The Oligosaccharides*, Ernest, I. and Hebky, J. eds, Academic Press, London.
3. Atwood, J. L., Davies, J. E. D., MacNicol, D. D. and Vögtle, F. eds (1996). *Comprehensive Supramolecular Chemistry*, vol. 3, Pergamon, Oxford.
4. Whistler, R. L. and BeMiller, J. N. (1992). *Industrial Gums. Polysaccharides and Their Derivatives*, Academic Press, London.
5. Aspinall, G. O. ed. (1982, 1983, 1985). *The Polysaccharides*, vols. 1–3, Academic Press, London.
6. Dumitriu, S. ed. (1996). *Polysaccharides in Medicinal Applications*, Marcel Dekker, New York.
7. Tombs, M. P. and Harding, S. E. (1998). *An Introduction to Polysaccharide Biotechnology*, Taylor and Francis, London.
8. Dumitriu, S. ed. (1998). *Polysaccharides. Structural Diversity and Functional Versatility*, Marcel Dekker, New York.

9. Stone, B. A. and Svensson, B. Enzymic depolymerisation and synthesis of polysaccharides, in *Glycoscience — Chemistry and Chemical Biology*, Thiem, J., Fraser-Reid, B. O. and Tatsuta, K. T. eds, vol. 3, Springer-Verlag.

10. Whistler, R. L. and Paschall, E. F. (1965). *Starch: Chemistry and Technology*, vol. 1, Academic Press, London.

11. Stone, B. A. (1996). Cereal grain carbohydrates, in *Cereal Grain Quality*, Henry, R. J. and Kettlewell, P. S. eds, Chapman and Hall, London, p. 251.

12. Tomasik, P. and Schilling, C. H. (1998). *Adv. Carbohydr. Chem. Biochem.*, **53**, 263.

13. Haigler, C. H. and Weimer, P. J. eds (1991). *Biosynthesis and Biodegradation of Cellulose*, Marcel Dekker, New York.

14. Kennedy, J. F., Phillips, G. O., Williams, P. A. and Piculell, L. eds (1995). *Cellulose and Cellulose Derivatives: Physico-chemical Aspects and Industrial Applications*, Woodhead Publishing, Cambridge.

15. Klemm, D., Philipp, B., Heinze, T., Heinze, U. and Wagenknecht, W. (1998). *Comprehensive Cellulose Chemistry*, vol. I. *Fundamentals and Analytical Methods*, Wiley, Chichester.

16. Fierobe, H.-P., Pagès, S., Bélaïch, A., Champ, S., Lexa, D. and Bélaïch, J.-P. (1999). *Biochemistry*, **38**, 12822.

17. Carpita, N. and Vergara, C. (1998). *Science*, **279**, 672.

18. Foster, A. B. and Webber, J. M. (1960). *Adv. Carbohydr. Chem. Biochem.*, **15**, 371.

19. Muzzarelli, R. A. A. (1977). *Chitin*, Pergamon Press, Oxford.

20. Muzzarelli, R. A. A. and Peter, M. G. eds (1997). *Chitin Handbook*, European Chitin Society, ISBN 88-86889-01-1.

21. Jolles, P. and Muzzarelli, R. A. A. eds (1999). *Chitin and Chitinases*, Birkhauser Verlag, Basel.

22. Kobayashi, S., Kiyosada, T. and Shoda, S.-i. (1996). *J. Am. Chem. Soc.*, **118**, 13113.

23. Comper, W. D. (1981). *Heparin (and Related Polysaccharides). Structural and Functional Properties*, Gordon and Breach, London.

24. Casu, B. (1985). *Adv. Carbohydr. Chem. Biochem.*, **43**, 51.

25. Alban, S. (1997). Carbohydrates with anticoagulant and antithrombotic properties, in *Carbohydrates in Drug Design*, Witczak, Z. J. and Nieforth, K. A. eds, Marcel Dekker, New York, p. 209.

26. Linhardt, R. J. and Toida, T. (1997). Heparin oligosaccharides: new analogues development and applications, in *Carbohydrates in Drug Design*, Witczak, Z. J. and Nieforth, K. A. eds, Marcel Dekker, New York, p. 277.

27. Conrad, H. E. (1998). *Heparin-binding Proteins*, Academic Press, London.

28. Banner, D. W. (2000). *Nature*, **404**, 449.

29. Petitou, M., Hérault, J.-P., Bernat, A., Driguez, P.-A., Duchaussoy, P., Lormeau, J.-C. and Herbert, J.-M. (1999). *Nature*, **398**, 417.

30. Kochetkov, N. K. and Knirel, Yu. A. (1989). *Sov. Sci. Rev. B. Chem.*, **13**, 1.

31. Shibaev, V. N. (1986). *Adv. Carbohydr. Chem. Biochem.*, **44**, 277.

32. Aspinall, G. O., Chatterjee, D. and Brennan, P. J. (1995). *Adv. Carbohydr. Chem. Biochem.*, **51**, 169.

33. Zähringer, U., Lindner, B. and Rietschel, E. Th. (1994). *Adv. Carbohydr. Chem. Biochem.*, **50**, 211.

34. Jennings, H. J. (1983). *Adv. Carbohydr. Chem. Biochem.*, **41**, 155.

35. Lindberg, B. (1990). *Adv. Carbohydr. Chem. Biochem.*, **48**, 279.

36. Unger, F. M. (1981). *Adv. Carbohydr. Chem. Biochem.*, **38**, 323.

Chapter 11

Glycoconjugates and Glycobiology[1-4]

Glycoconjugates

D-Glucose, as an early entrant into the glycolysis sequence, is crucial to the well being of virtually all organisms. This hexose, together with D-fructose and D-galactose, make up most of the disaccharides that are found free in Nature. If one adds in L-arabinose, D-xylose, D-mannose, D-glucosamine and various uronic acids, one now has most of the constituents of the various polysaccharides. There exists, however, another class of biopolymer, often largely composed of carbohydrate and including other monosaccharides such as D-galactosamine, L-fucose and *N*-acetylneuraminic acid, where the oligosaccharide or polysaccharide is not found free but is joined chemically to another biomolecule — the so-called *glycoconjugates*.[5-12] We have seen one example of this class already — the blood-group oligosaccharides joined to a lipid or protein. The importance of these glycoconjugates has been recognized only in the last half century and their prominence established in the last quarter century; a brief comment about the more important types follows.

Glycoproteins (Proteoglycans)[13]

Glycoproteins, as the name implies, consist of a saccharide glycosidically linked to a protein. *Proteoglycans*, in which the glycan is the major component of the molecule, are a subclass of the glycoproteins and the linear polysaccharide often contains amino sugars (glycosaminoglycans). Some glycoproteins contain only a single carbohydrate chain; other proteoglycans are virtually "enveloped" in carbohydrate from multiple (as few as three) attachments.

Glycoproteins are divided into two groups that are differentiated by the type of linkage between the carbohydrate and the protein:

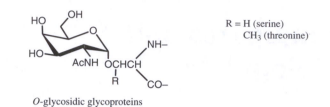

R = H (serine)
CH₃ (threonine)

O-glycosidic glycoproteins

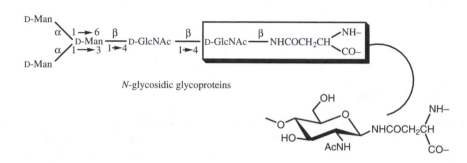

N-glycosidic glycoproteins

Starting from these two "bridgehead" structures, which are common in the *O*-glycosidic and invariant in the *N*-glycosidic glycoproteins, the addition of further monosaccharide residues builds up the various heterosaccharides found in the glycoproteins.

Glycolipids

The *glycolipids* are an incredibly diverse and structurally complex set of biomolecules in which the carbohydrate portion is often attached to a diglyceride or ceramide:

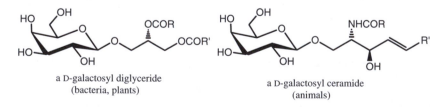

a D-galactosyl diglyceride
(bacteria, plants)

a D-galactosyl ceramide
(animals)

In the more complex glycolipids, such as the gangliosides (one particular member of the glycosphingolipids[14,15]), the D-galactose in the ceramide is replaced by D-glucose and the resulting oligosaccharide is adorned by one or more *N*-acetylneuraminic acid residues.

In just the last decade, the structure and function of some truly remarkable glycolipids have been unravelled, namely the "GPI anchors", which not only attach proteins to the plasma membrane but also are responsible for the transport of glycoproteins across membranes.

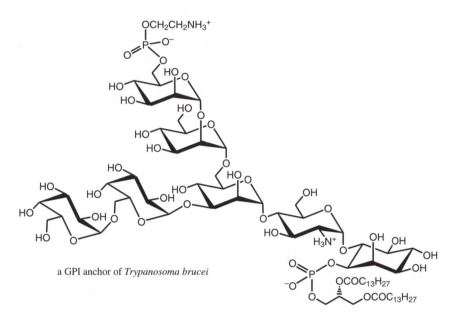

a GPI anchor of *Trypanosoma brucei*

Lipopolysaccharides[16]

Lipopolysaccharides consist of (usually) a polysaccharide glycosidically linked to a lipid, for example, lipid A:

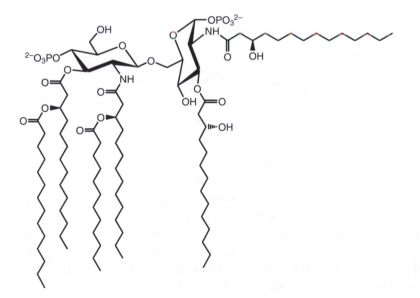

In light of the array of complex structures just mentioned, why has Nature gone to such trouble to construct carbohydrates with a protein attached, proteins virtually covered in carbohydrate, carbohydrates with fatty ends and fats with a bound carbohydrate? Obviously, but only delineated recently, there is a role for these glycoconjugates in Nature — the study of the role of carbohydrates in the world of biology is coined, *glycobiology*.[17-27] Because of the burgeoning mass of material in this rapidly developing field, only two topical examples will be mentioned, namely that of heparin and sialyl Lewis[x].

Glycobiology

Heparin

Mention was made in the last chapter of a small polysaccharide that was able to mimic the action of the anticoagulant, heparin, and some further discussion is warranted of this successful piece of glycobiological investigation.

Some years ago, van Boeckel and Petitou, among others, were able to identify and synthesize a pentasaccharide which constituted the antithrombin binding site of heparin:[28]

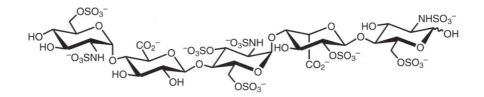

Although this pentasaccharide represented a potential antithrombotic drug, it lacked the other elements of heparin that are involved in the attraction of thrombin to the bound antithrombin. To obtain a more potent and well tolerated drug, it was decided to synthesize longer oligosaccharides.

From knowledge of the mechanism of action of heparin, it appeared that any oligosaccharide should be composed of an antithrombin binding domain (the pentasaccharide A-domain) coupled to a thrombin binding domain (T-domain). In the simplest construct, the two domains were kept identical:

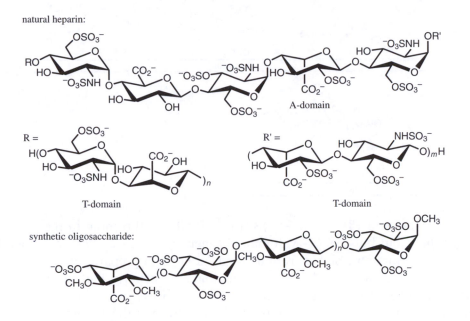

With $n = 9$, the "20-mer" oligosaccharide exhibited an antithrombotic activity similar to that for natural heparin but still displayed most of the unwanted side effects of heparin. Could it be that the synthetic molecule presented an efficient A-domain but needed a differently constructed T-domain to avoid the side effects? If so, should the T-domain be at the reducing or the non-reducing end of the A-domain — molecular modeling and crystallography studies suggested the latter:

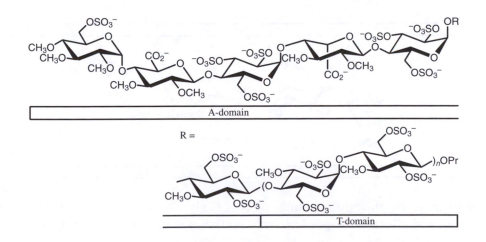

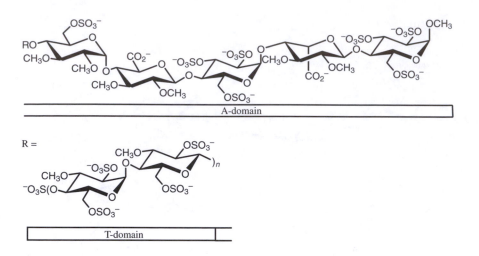

The "A-T-domain 18-mer" ($n = 6$) scarcely inhibited thrombin in the presence of antithrombin but the "T-A-domain 19-mer" ($n = 7$) was "as potent as the most active fraction isolated from a standard heparin preparation".[29,30]

Although an apparently excellent analogue of heparin had been designed and synthesized, the interaction was still not specific to just thrombin. It was thought that the problem was the polyanionic nature of the T-domain. In order to overcome this deficiency, a neutral linker was employed which eventually realized an oligosaccharide with the potential to become the first specific, antithrombotic drug:[31]

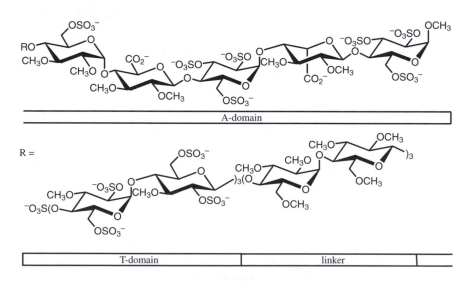

The success of this outstanding piece of glycobiology surely justifies the monumental efforts in oligosaccharide synthesis over the past two decades.

Sialyl Lewis[x] [22,32–35]

A sialic acid, *N*-acetylneuraminic acid (5-acetamido-3,5-dideoxy-D-*glycero*-D-*galacto*-2-nonulosonic acid) is a common terminal sugar found on the surface glycoprotein of many mammalian (*e.g.* leucocytes), bacterial and cancer cells. One of the most studied oligosaccharides containing this sialic acid is *sialyl Lewis[x]*:

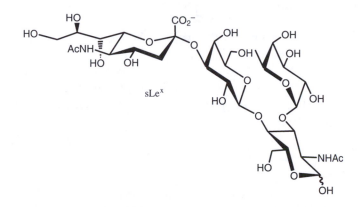

Just a decade ago, a new class of carbohydrate-binding glycoprotein, termed *selectins*, was found on the surface of certain cells and designated E-(endothelium), P-(platelet) and S-selectin (lymphocyte). Each of these selectins recognizes and binds, to varying degrees, sialyl Lewis[x] and it is the E-selectin interaction which occurs at the very early stages of the inflammatory reaction.

When tissue injury occurs, the endothelial cells respond by synthesizing increased levels of E-selectin to recruit neutrophils from the circulating blood to the site of injury. This process involves an adhesion that is governed by multivalent interactions between E-selectin and sialyl Lewis[x]. When too many neutrophils are recruited, there is the danger of damage to normal cells and the risk of inflammation. Therefore, with such a detailed understanding of this whole piece of glycobiology, there has been an enormous effort to produce sialyl Lewis[x] "mimics" (glycomimetics) to prevent the adhesion process and so lead to anti-inflammatory agents useful in the treatment of diseases such as asthma and arthritis. There is also evidence to suggest that a similar approach could arrest the process of metastasis and so lead to an effective treatment for cancer.

The synthesis of sialyl Lewis[x] by any number of academic and industrial laboratories has followed both traditional chemical routes and enzyme assisted processes[36–40] — the latter approach has yielded amounts in excess of a kilogram! These successes have allowed for detailed structure–function relationships and a study of the conformation of sialyl Lewis[x] in both the free and bound forms. A consequence of the latter has been to design and produce mimics of sialyl Lewis[x].[41,42] Wong, for instance, has synthesized molecules with a much improved binding to E-selectin over the natural ligand, sialyl Lewis^x:

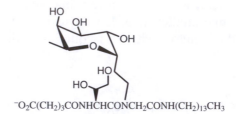

$$^-O_2C(CH_2)_3CONHCHCONCH_2CONH(CH_2)_{13}CH_3$$

These synthetic molecules are often non-carbohydrate in nature and, hence, show good stability with an associated and improved oral tolerance. Finally, any glycomimetic that can be prepared in a "multivalent" form generally exhibits improved activity.[43,44]

References

1. Kováč, P. ed. (1994). *Synthetic Oligosaccharides: Indispensable Probes for the Life Sciences*, ACS Symposium Series 560, American Chemical Society, Washington, DC.
2. Lehmann, J. (1998) (translated into English by Haines, A. H.). *Carbohydrates: Structure and Biology*, Thieme, Stuttgart.
3. Duus, J. O., St. Hilaire, P. M., Meldal, M. and Bock, K. (1999). *Pure Appl. Chem.*, **71**, 755.
4. Sears, P. and Wong, C.-H. (1999). *Angew. Chem. Int. Ed.*, **38**, 2300.
5. Horowitz, M. I. and Pigman, W. eds (1977, 1978). *The Glycoconjugates*, vols. I, II, Academic Press, London.
6. Horowitz, M. I., ed. (1982). *The Glycoconjugates*, vols. III, IV, Academic Press, London.
7. Roberts, D. D. and Mecham, R. P., eds (1993). *Cell Surface and Extracellular Glycoconjugates: Structure and Function*, Academic Press, London.
8. Lee, Y. C. and Lee, R. T. eds (1994). *Neoglycoconjugates: Preparation and Applications*, Academic Press, London.
9. Davis, B. G. (1999). *J. Chem. Soc., Perkin Trans. 1*, 3215.
10. Arsequell, G. and Valencia, G. (1999). *Tetrahedron*: Asymmetry, **10**, 3045.
11. Marcaurelle, L. A. and Bertozzi, C. R. (1999). *Chem. Eur. J.*, **5**, 1384.
12. St Hilaire, P. M. and Meldal, M. (2000). *Angew. Chem. Int. Ed.*, **39**, 1162.
13. Hounsell, E. F. ed. (1993). *Glycoprotein Analysis in Biomedicine, Methods in Molecular Biology*, Walker, J. M. ed., vol. 14, Humana Press, Clifton NJ.
14. Kolter, T. and Sandhoff, K. (1999). *Angew. Chem. Int. Ed.*, **38**, 1532.
15. Vankar, Y. D. and Schmidt, R. R. (2000). *Chem. Soc. Rev.*, **29**, 201.
16. Anderson, L. and Unger, F. M. eds (1983). *Bacterial Lipopolysaccharides: Structure, Synthesis and Biological Activities*, ACS Symposium Series 231, American Chemical Society, Washington, DC.
17. Karlsson, K.-A. (1991). *Trends Pharmacol. Sci.*, **12**, 265.
18. Rudd, P. M. and Dwek, R. A. (1991). *Chem. Ind.*, 660.
19. Varki, A. (1993). *Glycobiology*, **3**, 97.
20. Fukuda, M. and Kobata, A. eds (1993). *Glycobiology: a Practical Approach*, Oxford University Press Inc, New York.

21. Kobata, A. (1993). *Acc. Chem. Res.*, **26**, 319.
22. Fukuda, M. and Hindsgaul, O. eds (1994). *Molecular Glycobiology, Frontiers in Molecular Biology*, Hames, B. D. and Glover, D. M. eds, Oxford University Press Inc, New York.
23. Dwek, R. A. (1995). *Biochem. Soc. Trans.*, **23**, 1.
24. Dwek, R. A. (1996). *Chem. Rev.*, **96**, 683.
25. Townsend, R. R. and Hotchkiss, A. T., Jr. eds (1997). *Techniques in Glycobiology*, Marcel Dekker, New York.
26. Witczak, Z. J. and Nieforth, K. A. eds (1997). *Carbohydrates in Drug Design*, Marcel Dekker, New York.
27. Varki, A., Cummings, R., Esko, J., Freeze, H., Hart, G. and Marth, J. eds (1999). *Essentials of Glycobiology*, CSHL Press, New York.
28. van Boeckel, C. A. A. and Petitou, M. (1993). *Angew. Chem. Int. Ed. Engl.*, **32**, 1671.
29. Petitou, M., Hérault, J.-P., Bernat, A., Driguez, P.-A., Duchaussoy, P., Lormeau, J.-C. and Herbert, J.-M. (1999). *Nature*, **398**, 417.
30. Sinaÿ, P. (1999). *Nature*, **398**, 377.
31. Driguez, P.-A., Lederman, I., Strassel, J.-M., Herbert, J.-M. and Petitou, M. (1999). *J. Org. Chem.*, **64**, 9512.
32. Sears, P. and Wong, C.-H. (1996). *Proc. Natl. Acad. Sci. USA*, **93**, 12086.
33. Sears, P. and Wong, C.-H. (1998). *Chem. Commun.*, 1161.
34. Simanek, E. E., McGarvey, G. J., Jablonowski, J. A. and Wong, C.-H. (1998). *Chem. Rev.*, **98**, 833.
35. Wong, C.-H. (1999). *Acc. Chem. Res.*, **32**, 376.
36. Yoshida, M., Kiyoi, T., Tsukida, T. and Kondo, H. (1998). *J. Carbohydr. Chem.*, **17**, 673.
37. Ellervik, U. and Magnusson, G. (1998). *J. Org. Chem.*, **63**, 9314.
38. Zhang, Y.-M., Brodzky, A. and Sinaÿ, P. (1998). *Tetrahedron*: Asymmetry, **9**, 2451.
39. Blixt, O. and Norberg, T. (1998). *J. Org. Chem.*, **63**, 2705.
40. Komba, S., Galustian, C., Ishida, H., Feizi, T., Kannagi, R. and Kiso, M. (1999). *Angew. Chem. Int. Ed.*, **38**, 1131.
41. Kretzschmar, G. (1998). *Tetrahedron*, **54**, 3765.
42. Ellervik, U., Grundberg, H. and Magnusson, G. (1998). *J. Org. Chem.*, **63**, 9323.
43. Roy, R. (1998). The chemistry of neoglycoconjugates, in *Carbohydrate Chemistry*, Boons, G.-J. ed., Blackie, Edinburgh, p. 243.
44. Roy, R. (1999). *Carbohydrates in Europe*, Dec. 27, p. 34.

Chapter 12

Carbohydrate-based Vaccines[1-6]

The process of immunization probably began in the late eighteenth century when Edward Jenner (1749–1823), a medical practitioner, noticed that milkmaids, who presented the external symptoms (lesions and vesicles) of cowpox, never succumbed to the more deadly smallpox. Jenner removed some of the fluid from such a cowpox vesicle and injected it into a young boy, thereby immunizing him against the ravages of smallpox. In doing so, Jenner had caused the boy's immune system to raise antibodies against the (glyco)protein coat of the cowpox and these antibodies were sufficiently broad in their mode of action as to be able to neutralize any invading smallpox virus.

Since these early experiments, the design and production of vaccines has become instrumental in the prevention of disease; global immunization programmes have eradicated smallpox and control polio, meningitis and tuberculosis. It has been estimated that seventy per cent of the vaccines in use today are based on carbohydrate antigens. How, then, are these vaccines produced?

In the early years, vaccines were based on whole-organism preparations or on toxins isolated from different bacteria. Later, when a better understanding of the structure and chemical constitution of bacteria was obtained, vaccines were produced so that type-specific protection against a particular infection was possible. To quote Jennings:[7]

> *The concept of using a purified polysaccharide immunogen devoid of its accompanying, complex, bacterial mass is technically elegant.*

Another way of producing a desired polysaccharide antigen is by chemical synthesis. In addition, could this approach be widened to include not just bacterial infections but some of the most common and more deadly cancers? What follows is such a "case study".

The globo-H tumour antigen has been identified on the surface of both prostate- and breast-cancer cells:

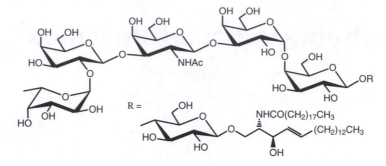

In an effort to develop a vaccine that would be active in an adjuvant or minimum-disease setting, *viz.* after excision or shrinkage of the original tumour, Danishefsky and coworkers first prepared the active antigen by chemical synthesis:[8]

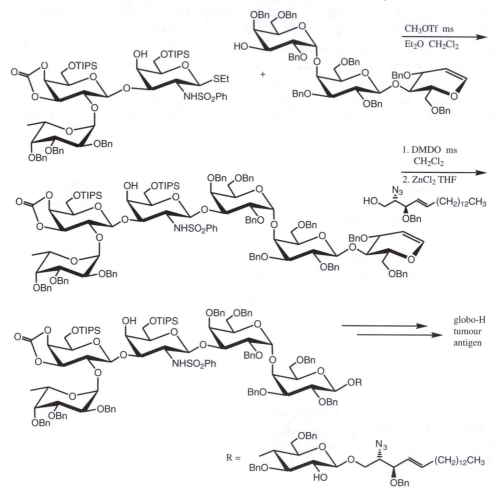

In a spectacular application of the "glycal assembly" method, two trisaccharides were first assembled, then joined and the ceramide aglycon finally added. In order to obtain an antigen that was immunogenically competent, the synthesis was modified to produce the allyl glycoside:

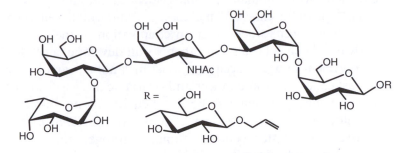

Chemical manipulation of the allyl group, *via* ozonolysis and reductive amination with the protein, keyhole limpet haemocyanin, then allowed the preparation of the desired glycoconjugate in a polyvalent form:

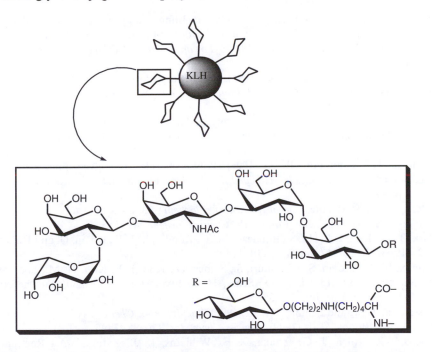

When the above synthetic vaccine was administered to five patients with progressive and recurrent prostate cancer, there was a high and specific immune response — such a response was polyclonal in nature but focussed against various portions of the globo-H antigen. Just as importantly, the raised sera were able to recognize the globo-H antigen in its natural context, *viz.* on the surface of prostate cancer cells, and cause cell lysis.[9] The globo-H antigen has also been prepared by

Schmidt (TCA methodology)[10] and by Boons (thioglycoside and glycosyl fluoride methodology).[11]

In further developments, Danishefsky has reported the synthesis of the 4-pentenyl glycoside of the globo-H tumour antigen,[12] as well as the carbohydrate antigens associated with human colonic adenocarcinoma cells,[13] gastric adenocarcinoma,[14] prostate cancer,[15] melanoma[16] and small cell lung carcinoma.[17] In some cases, vaccines derived from conjugation of these antigens to a carrier are in advanced clinical trials and others await development.[18] All in all, it is hoped that polyvalent antigen vaccines against melanoma, sarcoma, small cell lung cancer, breast cancer, prostate cancer and ovarian cancer will soon exist.[16,19]

A similar synthetic approach has been taken by Wong[20] and Pozsgay[21] towards the development of vaccines for melanoma and for *Shigella dysenteriae* type 1, respectively, the latter being a human pathogen that is the major causative organism of endemic and epidemic dysentery worldwide.

References

1. Bishop, C. T. and Jennings, H. J. (1982). Immunology of polysaccharides, in *The Polysaccharide*, Aspinall, G. O. ed., vol. 1, Academic Press, London, p. 291.
2. Bell, R. and Torrigiani, G. eds (1987). *Towards Better Carbohydrate Vaccines*, Wiley, Chichester.
3. Jennings, H. J. and Sood, R. K. (1994). Synthetic glycoconjugates as human vaccines, in *Neoglycoconjugates: Preparation and Applications*, Lee, Y. C. and Lee, R. T. eds, Academic Press, London, p. 325.
4. Lee, C.-J. (1996). Bacterial capsular polysaccharides: immunogenicity and vaccines, in *Polysaccharides in Medicinal Applications*, Dumitriu, S. ed., Marcel Dekker, New York, p. 411.
5. Jennings, H. J. and Pon, R. A. (1996). Polysaccharides and glycoconjugates as human vaccines, in *Polysaccharides in Medicinal Applications*, Dumitriu, S. ed., Marcel Dekker, New York, p. 443.
6. Pazur, J. H. (1998). *Adv. Carbohydr. Chem. Biochem.*, **53**, 201.
7. Jennings, H. J. (1983). *Adv. Carbohydr. Chem. Biochem.*, **41**, 155.
8. Park, T. K., Kim, I. J., Hu, S., Bilodeau, M. T., Randolph, J. T., Kwon, O. and Danishefsky, S. J. (1996). *J. Am. Chem. Soc.*, **118**, 11488.
9. Ragupathi, G., Slovin, S. F., Adluri, S., Sames, D., Kim, I. J., Kim, H. M., Spassova, M., Bornmann, W. G., Lloyd, K. O., Scher, H. I., Livingston, P. O. and Danishefsky, S. J. (1999). *Angew. Chem. Int. Ed.*, **38**, 563.
10. Lassaletta, J. M. and Schmidt, R. R. (1996). *Liebigs Ann. Chem.*, 1417.
11. Zhu, T. and Boons, G.-J. (1999). *Angew. Chem. Int. Ed.*, **38**, 3495.
12. Allen, J. R., Allen, J. G., Zhang, X.-F., Williams, L. J., Zatorski, A., Ragupathi, G., Livingston, P. O. and Danishefsky, S. J. (2000). *Chem. Eur. J.*, **6**, 1366.
13. Deshpande, P. P., Kim, H. M., Zatorski, A., Park, T.-K., Ragupathi, G., Livingston, P. O., Live, D. and Danishefsky, S. J. (1998). *J. Am. Chem. Soc.*, **120**, 1600.
14. Chen, X.-T., Sames, D. and Danishefsky, S. J. (1998). *J. Am. Chem. Soc.*, **120**, 7760.
15. Kuduk, S. D., Schwarz, J. B., Chen, X.-T., Glunz, P. W., Sames, D., Ragupathi, G., Livingston, P. O. and Danishefsky, S. J. (1998). *J. Am. Chem. Soc.*, **120**, 12474.

16. Livingston, P., Ragupathi, G. and Danishefsky, S. J. (1998). *Abstract Book*, XIXth International Carbohydrate Symposium, San Diego.
17. Allen, J. R. and Danishefsky, S. J. (1999). *J. Am. Chem. Soc.*, **121**, 10875.
18. Koganty, R. R., Qiu, D., Gandhi, S. S., Reddish, M. A. and Longenecker, B. M. (1998). *Abstract Book*, XIXth International Carbohydrate Symposium, San Diego.
19. Danishefsky, S. J. and Allen, J. R. (2000). *Angew. Chem. Int. Ed.*, **39**, 837.
20. Sears, P. and Wong, C.-H. (1998). *Chem. Commun.*, 1161.
21. Pozsgay, V. (1998). *Angew. Chem. Int. Ed.*, **37**, 138.

Appendix

Carbohydrate Nomenclature

By now, the reader will have gained some idea of the basic rules of carbohydrate nomenclature. Fortunately, these rules have recently been reformulated and are readily available in several places as the "Nomenclature of carbohydrates":

Pure Appl. Chem., 1996, **68**, 1919.
Adv. Carbohydr. Chem. Biochem., 1997, **52**, 43.
Carbohydr. Res., 1997, **297**, 1.
J. Carbohydr. Chem., 1997, **16**, 1191.
www.chem.qmw.ac.uk/iupac/2carb/index.html

These new rules are not cast in stone and the use of older nomenclature may have advantages in certain cases.

The Literature of Carbohydrates

Reference Literature

Chemical Abstracts still remains the most important reference source for literature on carbohydrates, whether it be by casual scanning of "33-Carbohydrates" in each weekly issue or by a formal search utilizing the Author, General Subject, Chemical Substance or Formula Index. The Thirteenth Collective Index enables rapid searching up to 1996.

—*CA Selects* (http://www.cas.org/PRINTED/caselects.html) is a weekly publication that provides all the relevant abstracts in a certain subject area, *e.g.* Carbohydrates.

—*CAS Online* (http://www.cas.org/) is a very rapid, computer-based method of searching *Chemical Abstracts*.

—*SciFinder* (http://www.cas.org/SCIFINDER/) *Scholar* (http://www.cas.org/SCIFINDER/SCHOLAR/) uses the same database as *CAS Online* but offers different search features and really does bring the chemical literature onto one's desktop.

Beilsteins Handbuch der Organischen Chemie is an ingenious system that details individual classes of chemical compounds in a particular volume (Hauptwerk) and then updates the information with supplements — carbohydrates are to be found in volumes 17, 18, 19 and 31. The hard copy of Beilstein has essentially been replaced by *CrossFire*, an online version, in English (http://www.beilstein.com/beilst_2.shtml).

Rodd's Chemistry of Carbon Compounds provides another source of literature on carbohydrates, located in volumes I^F (1967) and I^G (1976) and their supplements (1983, to I^{FG} and 1993, to I^E/I^F/I^G).

Methoden der Organischen Chemie (Houben-Weyl) is another multi-volume work that describes, in vol. E14a/3, various aspects of the chemistry of carbohydrates and their derivatives.

Primary Literature

General papers on all aspects of the chemistry and biochemistry of carbohydrates now appear in primary journals and this is simply a reflection of the increased interest in carbohydrates shown by mainstream chemists and biochemists. There are various specialist journals devoted to the chemistry of carbohydrates, namely *Carbohydrate Research* (1965–), the *Journal of Carbohydrate Chemistry* (1982–), *Carbohydrate Letters* (1994–), *Carbohydrate Polymers* (1981–), *Glycobiology* (1990–) and *Glycoconjugate Journal* (1984–).

Monographs and Related Works

Methods in Carbohydrate Chemistry (1962–) is an excellent series that provides discussion, references and experimental procedures for a host of transformations in carbohydrates; volume II is probably one of the most valuable works ever published for carbohydrate chemists.

Advances in Carbohydrate Chemistry (1945–1968) and *Advances in Carbohydrate Chemistry and Biochemistry* (1969–) provide a set of excellent reviews on all aspects of carbohydrate chemistry and biochemistry.

Specialist Periodical Reports, Carbohydrate Chemistry (1968–), somewhat quaintly known as "the red book", is an annual review of the carbohydrate literature that is compiled by a team of reviewers.

The Monosaccharides (1963), by Jaroslav Staněk and co-authors, is a bible for carbohydrate chemists. The four volume treatise, *The Carbohydrates, Chemistry and Biochemistry*, edited by Ward Pigman and Derek Horton, is again a "must" for all researchers in carbohydrates — a new edition is promised for publication in 2000 as the second edition (vol. IA, 1972; vol. IB, 1980; vols. IIA and IIB, 1970) is really showing its age.

Carbohydrates — a Source Book, edited by Peter M. Collins, can be of use if one wishes to consult an "encyclopedia" of individual carbohydrate compounds that lists the relevant physical data and gives references for methods of preparation.

Comprehensive Natural Products Chemistry has just published (1999) a whole volume (vol. 3, "Carbohydrates and their derivatives including tannins, cellulose, and related lignins") devoted to various aspects of carbohydrates.

Recent Edited Works

A popular and recent trend in many areas of science has been the publication of works that emanate either from a conference or from the desire of one person (the editor) to present recent progress in a certain field. As such, these works contain articles by many different authors and there are the obvious problems associated with differing styles and presentation.

Modern Methods in Carbohydrate Synthesis (eds. Khan, S. H. and O'Neill, R. A.; Harwood Academic, Netherlands, 1996) was the first of these modern edited works and contains many useful articles on all aspects of carbohydrate chemistry.

Preparative Carbohydrate Chemistry (ed. Hanessian, S.; Marcel Dekker, New York, 1997) again contains many chapters, some more up to date than others, and a multi-chapter section on unpublished aspects of the "remote activation" concept. A novel aspect of the book is the inclusion of experimental details for the theme reaction(s) of each chapter.

Carbohydrate Chemistry (ed. Boons, G.-J.; Blackie, Edinburgh, 1998) contains a wealth of information about carbohydrates and has powerful chapters dealing with the synthesis of glycosides and the chemistry of neoglycoconjugates.

Bioorganic Chemistry: Carbohydrates (ed. Hecht, S. M.; Oxford University Press, Oxford, 1998) is the final of three volumes on bioorganic chemistry and is unique in that the thirteen chapters are designed to fit into a one semester course.

Carbohydrate Mimics: Concepts and Methods (ed. Chapleur, Y.; Wiley-VCH, Weinheim, 1998) is a compilation of works from laboratories around the world that describes the synthesis of carbohydrate mimics such as azasugars, *C*-linked sugars, carbasugars, aminocyclopentitols and carbocycles.

Recent Textbooks

As well as supplying the scientific community with the latest literature and review articles, it is also necessary to provide textbooks for use by undergraduates, postgraduates and young researchers. A textbook demands a certain style of the author, to present a goodly part of a subject in an easily understandable, friendly and readable manner.

Carbohydrate Chemistry: Monosaccharides and Their Oligomers by Hassan S. El Khadem (Academic Press, London, 1988) was just about the first of a rush of new-wave textbooks on carbohydrates. The emphasis is on monosaccharides, with only a handful of literature references.

Modern Carbohydrate Chemistry by Roger W. Binkley (Marcel Dekker, New York, 1988) gives a reasonable overview of monosaccharide chemistry with a more generous offering of literature references.

Monosaccharides: Their Chemistry and Their Roles in Natural Products by Peter M. Collins and Robert (Robin) J. Ferrier (Wiley, Chichester, 1995) is essentially the second edition of a small book [*Monosaccharide Chemistry* (Penguin, Harmondsworth, 1972)] that took the carbohydrate community by storm. The present book, written by two wise, wistful and knowledgeable carbohydrate chemists, is a mine of information, presumably gleaned from the years associated with "Specialist Periodical Reports, Carbohydrate Chemistry". You will not find any carbohydrate laboratory in the world without a copy of this gem.

Carbohydrates: Structure and Biology by Jochen Lehmann was originally published in German [*Kohlenhydrate. Chemie und Biologie* (Thieme, Stuttgart, 1996)] and subsequently translated by Alan H. Haines (Thieme, Stuttgart, 1998); the English version does not contain the chapter, "Chemical Aspects", which is in the German version. Chapter 2 of the English version, entitled "Biological Aspects", is an excellent summary of the role of carbohydrates in biology (glycobiology). This book is another "must" for any self-respecting carbohydrate chemist and represents excellent value for money.

Essentials of Carbohydrate Chemistry by John F. Robyt (Springer, New York, 1998) is another book with a biological emphasis; there is a heavy accent on aspects of the chemistry of sucrose throughout the book.

The Molecular and Supramolecular Chemistry of Carbohydrates by Serge David (Oxford University Press, Oxford, 1998) provides an overview of the physical, chemical and biological properties of carbohydrates. With eighteen chapters (and 320 pages), the book is wide ranging in its coverage.

Essentials of Carbohydrate Chemistry and Biochemistry by Thisbe K. Lindhorst (Wiley-VCH, Weinheim, 2000) is the latest textbook on carbohydrates. Somewhat unfortunately, no literature references are included.

Other Works

Carbohydrate Building Blocks by Mikael Bols (Wiley, Chichester, 1996) spends some sixty pages in discussing the chemistry of carbohydrates and then finishes with the main thrust of the book, a compendium of "building blocks" derived from carbohydrates for the synthesis of natural products.

Monosaccharide Sugars: Chemical Synthesis by Chain Elongation, Degradation, and Epimerization by Zoltán Györgydeák and István F. Pelyvás (Academic Press, London, 1998) describes a myriad of reactions for the elongation, degradation and isomerization of monosaccharides. A highlight of the book is the inclusion of detailed experimental descriptions of many of the transformations discussed.

Index

Note: Tables and Figures are indicated by *italic page numbers*, footnotes by suffix (n)